Hydraulics in Civil and Environmental Engineering

Hydraulics in Civil and Environmental Engineering

– Author –
Naseer Khan

Wordpen Academics
Publishing Science That Matter

ISBN: 978-93-90438-77-8 (Hardback)

Published by : BIO-GREEN BOOKS
4736/23, Ansari Road, Darya Ganj,
New Delhi – 110 002
Phone: 011-23245578, 23244987
E-mail: biogreenbooks@gmail.com

Printed at : Replika Press Pvt. Ltd.

Preface

The skills of environmental engineers are becoming increasingly important as we attempt to protect the fragile resources of our planet. Hydraulic analyses are a significant component of water resources studies and the design of hydraulic engineering works. They serve to assess and optimize flow phenomena and their interaction with structures and are also required to determine the dimensions of structures such as dams, weirs, waterways, and stilling basins as well as for the design of pumps and turbines. They define pressures and other required loads for stability analysis of the structures. Water resources engineering is the quantitative study of the hydrologic cycle — the distribution and circulation of water linking the earth's atmosphere, land and oceans. Surface runoff is measured as the difference between precipitation and abstractions, such as infiltration (which replenishes groundwater flow), surface storage and evaporation. Applications include the management of the urban water supply, the design of urban storm-sewer systems, and flood forecasting. Hydraulic engineering consists of the application of fluid mechanics to water flowing in an isolated environment (pipe, pump) or in an open channel (river, lake, ocean). Civil engineers are primarily concerned with open channel flow, which is governed by the interdependent interaction between the water and the channel. Applications include the design of hydraulic structures, such as sewage conduits, dams and breakwaters, the management of waterways, such as erosion protection and flood protection, and environmental management, such as prediction of the mixing and transport of pollutants in surface water. Hydroelectric-power development, water supply, irrigation and navigation are some familiar applications of water resources engineering involving the utilization of water for beneficial purposes. More recently, concern for preserving our natural environment and meeting the needs of developing countries has increased the importance of water resources engineering. Civil engineers play a vital role in the optimal planning, design and operation of water resource systems.

Common topics of design for hydraulic engineers include hydraulic structures such as dams, levees, water distribution networks, water collection networks, sewage collection networks, storm water management, sediment transport, and various other topics related to transportation engineering and geotechnical engineering. Equations developed from the principles of fluid dynamics and fluid mechanics are widely utilized by other engineering disciplines such as mechanical, aeronautical and even traffic engineers. Related branches include hydrology and rheology while related applications include hydraulic modelling, flood mapping, catchment flood management plans, shoreline management plans, estuarine strategies, coastal protection, and flood alleviation. Many respects the fundamentals of hydraulic engineering haven't changed since ancient times. Liquids are still moved for the most part by gravity through systems of canals and aqueducts, though the supply reservoirs may now be filled using pumps. The need for water has steadily increased from ancient times and the role of the hydraulic engineer is a critical one in supplying it. For example, without the efforts of people like William Mulholland the Los Angeles area would not have been able to grow as it has because it simply doesn't have enough local water to support its population. The modern hydraulic engineer uses the same kinds of computer-aided design (CAD) tools as many of the other engineering disciplines while also making use of technologies like computational fluid dynamics to perform the calculations to accurately predict flow characteristics, GPS mapping to assist in locating the best paths for installing a system and laser-based surveying tools to aid in the actual construction of a system.

The book will be an indispensable source for all professionals, researchers and students in this subject and for anyone working in the related areas for acquiring an up-to-date overviews.

—Naseer Khan

Contents

Preface *v*

1. **Introduction**

Hydraulics • Hydraulic Engineering • Fluid Dynamics • Magnetohydrodynamics • Boussinesq Approximation (Buoyancy)

2. **Lubrication Theory** **52**

Hele-Shaw Flow • Slender-Body Theory • Boussinesq Approximation (Water Waves) • Shallow Water Equations • Darcy's Law • Balanced Flow • Fluid

3. **Pipe Flow** **88**

Hagen–Poiseuille Equation • Moody Chart • Open-Channel Flow • Surge Control • Hydraulic Machinery • Automatic Transmission • Hydraulic Brake • High-Density Solids Pump

4. **Hydraulic Accumulator** **143**

Wallasey Great Float • Hydraulic Cylinder • Hydraulic Ram • Hydraulic Fluid • Hydraulic Circuit • Hydraulic Motor

5. **Hydraulic Power Network** **166**

Hydraulic Press • Hydraulic Rescue Tools • Wave • Flood • Principal Types and Causes

6. **Dimensional Analysis** **198**

Great Principle of Similitude • Polynomials and Transcendental Functions • Position vs Displacement • Percentages and Derivatives • Dimensionless Concepts • Theoretical Physics

7. Mainstream Theories **220**

Pipeline Transport • Hydraulic Structure • Turbulence Modelling • Spalart–Allmaras Turbulence Model

8. Atmospheric Dispersion Modelling **255**

Breakwater (Structure) • Beach Nourishment • Clapotis • Iribarren Number • Hudson's Equation • Revetment • Sand Dune Stabilization • Sand Engine • Seawall

Bibliography 292

Index 295

Chapter 1

Introduction

Hydraulics

Hydraulics is a topic in applied science and engineering dealing with the mechanical properties of liquids. At a very basic level hydraulics is the liquid version of pneumatics. Fluid mechanics provides the theoretical foundation for hydraulics, which focuses on the engineering uses of fluid properties.

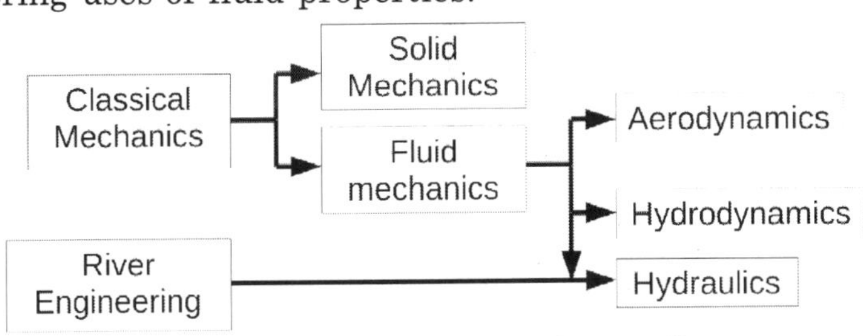

Figure: *Hydraulics and other studies*

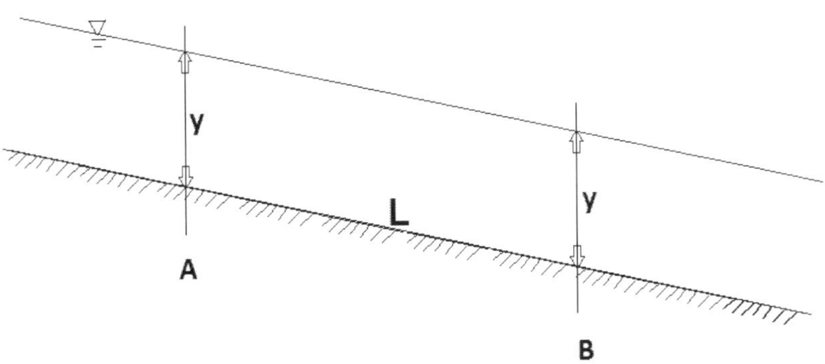

Figure: *An open channel, with a uniform depth, Open Channel Hydraulics deals with uniform and non-uniform streams.*

In fluid power, hydraulics is used for the generation, control, and transmission of power by the use of pressurized liquids. Hydraulic topics range through some part of science and most of engineering modules, and cover concepts such as pipe flow, dam design, fluidics and fluid control circuitry, pumps, turbines, hydropower, computational fluid dynamics, flow measurement, river channel behaviour and erosion.

Free surface hydraulics is the branch of hydraulics dealing with free surface flow, such as occurring in rivers, canals, lakes, estuaries and seas. Its sub-field open channel flow studies the flow in open channels.

The word "hydraulics" originates from the Greek word Qäñáõëéêüò (*hydraulikos*) which in turn originates from Uäùñ (*hydor*, Greek for water) and αPëüò (*aulos*, meaning pipe).

Ancient and Medieval Era

Early uses of water power date back to Mesopotamia and ancient Egypt, where irrigation has been used since the 6th millennium BC and water clocks had been used since the early 2nd millennium BC. Other early examples of water power include the Qanat system in ancient Persia and the Turpan water system in ancient China.

Greek / Hellenistic Age

The Greeks constructed sophisticated water and hydraulic power systems. An example is the construction by Eupalinos, under a public contract, of a watering channel for Samos, the Tunnel of Eupalinos. An early example of the usage of hydraulic wheel, probably the earliest in Europe, is the Perachora wheel (3rd century BC).

Notable is the construction of the first hydraulic automata by Ctesibius (flourished c. 270 BC) and Hero of Alexandria (c. 10 – 80 AD). Hero describes a number of working machines using hydraulic power, such as the force pump, which is known from many Roman sites as having been used for raising water and in fire engines.

China

In ancient China there was Sunshu Ao (6th century BC), Ximen Bao (5th century BC), Du Shi (circa 31 AD), Zhang Heng (78 - 139 AD), and Ma Jun (200 - 265 AD), while medieval China had Su Song (1020 - 1101 AD) and Shen Kuo (1031–1095). Du Shi employed a waterwheel to power the bellows of a blast furnace producing cast iron. Zhang Heng was the first to employ hydraulics to provide motive power in rotating an armillary sphere for astronomical observation.

Sri Lanka

In ancient Sri Lanka, hydraulics were widely used in the ancient kingdoms of Anuradhapura and Polonnaruwa. The discovery of the principle of the valve tower, or valve pit, for regulating the escape of water is credited to ingenuity more than 2,000 years ago. By the first century AD, several large-scale irrigation works had been completed. Macro- and micro-hydraulics to provide for domestic horticultural and agricultural needs, surface drainage and erosion control, ornamental and recreational water courses and retaining structures and also cooling systems were in place in Sigiriya, Sri Lanka. The coral on the massive rock at the site includes cisterns for collecting water.

Innovations in Ancient Rome

In Ancient Rome many different hydraulic applications were developed, including public water supplies, innumerable aqueducts, power using watermills and hydraulic mining. They were among the first to make use of the siphon to carry water across valleys, and used hushing on a large scale to prospect for and then extract metal ores. They used lead widely in plumbing systems for domestic and public supply, such as feeding thermae. Hydraulic mining was used in the gold-fields of northern Spain, which was conquered by Augustus in 25 BC. The alluvial gold-mine of Las Medulas was one of the largest of their mines. It was worked by at least 7 long aqueducts, and the water streams were used to erode the soft deposits, and then wash the tailings for the valuable gold content.

Modern era (c. 1600 – 1870)

Benedetto Castelli: In 1619 Benedetto Castelli (1576 - 1578–1643), a student of Galileo Galilei, published the book *Della Misura dell'Acque Correnti* or "On the Measurement of Running Waters", one of the foundations of modern hydrodynamics. He served as a chief consultant to the Pope on hydraulic projects, i.e., management of rivers in the Papal States, beginning in 1626.

Blaise Pascal

Blaise Pascal (1623–1662) studied fluid hydrodynamics and hydrostatics, centred on the principles of hydraulic fluids. His inventions include the hydraulic press, which multiplied a smaller force acting on a smaller area into the application of a larger force totaled over a larger area, transmitted through the same pressure (or same change of pressure) at both locations. Pascal's law or principle states that for an incompressible fluid at rest, the difference in pressure is proportional to the difference in height and this difference remains the same whether or not the overall pressure of the fluid is changed by applying an external force. This implies that by increasing the pressure at any point in a confined fluid, there is an equal increase at every other point in the container, i.e., any change in pressure applied at any point of the fluid is transmitted undiminished throughout the fluids.

Jean Louis Marie Poiseuille

A French physician, Poiseuille researched the flow of blood through the body and discovered an important law governing the rate of flow with the diameter of the tube in which flow occurred.

In the UK

Several cities developed city-wide hydraulic power networks in the 19th century, to operate machinery such as lifts, cranes, capstans and the like. Joseph Bramah was an early innovator and William Armstrong perfected the apparatus for power delivery on an industrial scale. In London, the London Hydraulic Power Company was a major supplier its pipes serving large parts of the West End of London, City and the Docks, but there were schemes restricted to single enterprises such as docks and railway goods yards.

Hydraulic Engineering

Hydraulic engineering as a sub-discipline of civil engineering is concerned with the flow and conveyance of fluids, principally water and sewage. One feature of these systems is the extensive use of gravity as the motive force to cause the movement of the fluids. This area of civil engineering is intimately related to the design of bridges, dams, channels, canals, and levees, and to both sanitary and environmental engineering.

Hydraulic engineering is the application of fluid mechanics principles to problems dealing with the collection, storage, control, transport, regulation, measurement, and use of water. Before beginning a hydraulic engineering project, one must figure out how much water is involved. The hydraulic engineer is concerned with the transport of sediment by the river, the interaction of the water with its alluvial boundary, and the occurrence of scour and deposition. "The hydraulic engineer actually develops conceptual designs for the various features which interact with water such as spillways and outlet works for dams, culverts for highways, canals and related structures for irrigation projects, and cooling-water facilities for thermal power plants."

Figure: *Hydraulic Flood Retention Basin (HFRB)*

Fundamental Principles

A few examples of the fundamental principles of hydraulic engineering include fluid mechanics, fluid flow, behaviour of real fluids, hydrology, pipelines, open channel hydraulics, mechanics of sediment transport, physical modelling, hydraulic machines, and drainage hydraulics.

Fluid Mechanics

Fundamentals of Hydraulic Engineering defines hydrostatics as the study of fluids at rest. Fluids at rest indicate that there exists a force, known as pressure, that acts upon its surroundings. This pressure, measured in N/m^2, is not constant throughout the body of fluid. Pressure, p, in a given body of fluid, increases with an increase in depth. Where the upward force on a body acts on the base and can be found by equation:

$$p = \rho g y$$

where,

ρ = density of water

g = specific gravity

y = depth of the body of liquid

Rearranging this equation gives you the pressure head $p/\rho g = y$. Four basic devices for pressure measurement are a piezometre, manometre, differential manometre, Bourdon gauge, as well as an inclined manometre.

As Prasuhn States: On undisturbed submerged bodies, pressure acts along all surfaces of a body in a liquid, causing equal perpendicular forces in the body to act against the pressure of the liquid. This reaction is known as equilibrium. More advanced applications of pressure are that on plane surfaces, curved surfaces, dams, and quadrant gates, just to name a few.

Behaviour of Real Fluids

Real and Ideal Fluids: The main difference between an ideal fluid and a real fluid is that for ideal flow $p_1 = p_2$ and for real flow $p_1 > p_2$. Ideal fluid is incompressible and has no viscosity. Real fluid has viscosity. Ideal fluid is only an imaginary fluid as all fluids that exist have some viscosity.

Viscous Flow

A viscous fluid will deform continuously under a shear force, whereas an ideal fluid doesn't deform.

Laminar Flow and Turbulence

The various effects of disturbance on a viscous flow are stable, transition and unstable.

Bernoulli's Equation

For an ideal fluid, Bernoulli's equation holds along streamlines.

$$p/\rho g + u^2/2g = p_1/\rho g + u_1^2/2g = p_2/\rho g + u_2^2/2g$$

Boundary Layer

Assuming a flow is bounded on one side only, and that a rectilinear flow passing over a stationary flat plate which lies parallel to the flow, the flow just upstream of the plate has a uniform velocity. As the flow comes into contact with the plate, the layer of fluid actually 'adheres' to a solid surface. There is then a considerable shearing action between the layer of fluid on the plate surface and the second layer of fluid. The second layer is therefore forced to decelerate (though it is not quite brought to rest), creating a shearing action with the third layer of fluid, and so on. As the fluid passes further along the plate, the zone in which shearing action occurs tends to spread further outwards. This zone is known as the 'boundary layer'. The flow outside the boundary layer is free of shear and viscous-related forces so it is assumed to act like an ideal fluid. "The intermolecular cohesive forces in a fluid are not great enough to hold fluid together. Hence a fluid will flow under the action of the slightest street and flow will continue as long as the stress is present. The flow inside the layer can be either viscous or turbulent, depending on Reynolds number.

Applications

Common topics of design for hydraulic engineers include hydraulic structures such as dams, levees, water distribution networks, water collection networks, sewage collection networks, storm water management, sediment transport, and various other topics related to transportation engineering and geotechnical engineering. Equations developed from the principles of fluid dynamics and fluid mechanics are widely utilised by other engineering disciplines such as mechanical, aeronautical and even traffic engineers.

Related branches include hydrology and rheology while related applications include hydraulic modelling, flood mapping, catchment flood management plans, shoreline management plans, estuarine strategies, coastal protection, and flood alleviation.

History

Earliest uses of hydraulic engineering were to irrigate crops and dates back to the Middle East and Africa. Controlling the movement and supply of water for growing food has been used for many thousands of years. One of the earliest hydraulic machines, the water clock was used in the early 2nd millennium BC. Other early examples of using gravity to move water include the Qanat system in ancient Persia and the very similar Turpan water system in ancient China as well as irrigation canals in Peru.

In ancient China, hydraulic engineering was highly developed, and engineers constructed massive canals with levees and dams to channel the flow of water for irrigation, as well as locks to allow ships to pass through. Sunshu Ao is considered the first Chinese hydraulic engineer. Another important Hydraulic Engineer in China, Ximen Bao was credited of starting the practice of large scale canal irrigation during the Warring States period (481 BC-221 BC), even today hydraulic engineers remain a respectable position in China. Before becoming President, Hu Jintao was a hydraulic engineer and holds an engineering degree from Tsinghua University

Eupalinos of Megara, was an ancient Greek engineer who built the Tunnel of Eupalinos on Samos in the 6th century BC, an important feat of both civil and hydraulic engineering. The civil engineering aspect of this tunnel was the fact that it was dug from both ends which required the diggers to maintain an accurate path so that the two tunnels met and that the entire effort maintained a sufficient slope to allow the water to flow.

Hydraulic engineering was highly developed in Europe under the aegis of the Roman Empire where it was especially applied to the construction and maintenance of aqueducts to supply water to and remove sewage from their cities. In addition to supplying the needs of their citizens they used hydraulic mining methods to prospect and extract alluvial gold deposits in a technique known as hushing, and applied the methods to other ores such as those of tin and lead.

Further advances in hydraulic engineering occurred in the Muslim world between the 8th to 16th centuries, during what is known as the Islamic Golden Age. Of particular importance was the 'water management technological complex' which was central to the Islamic Green Revolution and, by extension, a precondition for the emergence of modern technology. The various components of this 'toolkit' were developed in different parts of the Afro-Eurasian landmass, both within and beyond the Islamic world.

However, it was in the medieval Islamic lands where the technological complex was assembled and standardized, and subsequently diffused to the rest of the Old World. Under the rule of a single Islamic Caliphate, different regional hydraulic technologies were assembled into "an identifiable water management technological complex that was to have a global impact." The various components of this complex included canals, dams, the *qanat* system from Persia, regional water-lifting devices such as the *noria, shaduf* and screwpump from Egypt, and the windmill from Islamic Afghanistan. Other original Islamic developments included the *saqiya* with a flywheel effect from Islamic Spain, the reciprocating suction pump and crankshaft-connecting rod mechanism from Iraq, the geared and hydropowered water supply system from Syria, and the water purification methods of Islamic chemists.

Modern Times

In many respects the fundamentals of hydraulic engineering haven't changed since ancient times. Liquids are still moved for the most part by gravity through systems of canals and aqueducts, though the supply reservoirs may now be filled using pumps. The need for water has steadily increased from ancient times and the role of the hydraulic engineer is a critical one in supplying it. For example, without the efforts of people like William Mulholland the Los Angeles area would not have been able to grow as it has because it simply doesn't have enough local water to support its population. The same is true for many of our world's largest cities. In much the same way, the central valley of California could not have become such an important agricultural region without effective water management and distribution for irrigation. In a somewhat parallel way to what happened in California the creation of the Tennessee Valley Authority(TVA) brought work and prosperity to the South by building dams to generate cheap electricity and control flooding in the region, making rivers navigable and generally modernising life the region.

Leonardo da Vinci (1452–1519) performed experiments, investigated and speculated on waves and jets, eddies and streamlining. Isaac Newton (1642–1727) by formulating the laws of motion and his law of viscosity, in addition to developing the calculus, paved the way for many great developments in fluid mechanics. Using Newton's laws of motion, numerous 18th-century mathematicians solved many frictionless (zero-viscosity) flow problems. However, most flows are dominated by viscous effects, so engineers of the 17th and 18th centuries found the inviscid flow solutions unsuitable, and by experimentation they developed empirical equations, thus establishing the science of hydraulics.

Late in the 19th century, the importance of dimensionless numbers and their relationship to turbulence was recognised, and dimensional analysis was born. In 1904 Ludwig Prandtl published a key paper, proposing that the flow fields of low-viscosity fluids be divided into two zones, namely a thin, viscosity-dominated boundary layer near solid surfaces, and an effectively inviscid outer zone away from the boundaries. This concept explained many former paradoxes, and enabled subsequent engineers to analyze far more complex flows. However, we still have no complete theory for the nature of turbulence, and so modern fluid mechanics continues to be combination of experimental results and theory.

The modern hydraulic engineer uses the same kinds of computer-aided design (CAD) tools as many of the other engineering disciplines while also making use of technologies like computational fluid dynamics to perform the calculations to accurately predict flow characteristics, GPS mapping to assist in locating the best paths for installing a system and laser-based surveying tools to aid in the actual construction of a system.

Hydrostatics

Introduction: Hydrostatics is about the pressures exerted by a fluid at rest. Any fluid is meant, not just water. It is usually relegated to an early chapter in Fluid Mechanics texts, since its results are widely used in that study. The study yields many useful results of its own, however, such as forces on dams, buoyancy and hydraulic actuation, and is well worth studying for such practical reasons. It is an excellent example of deductive mathematical physics, one that can be understood easily and completely from a very few fundamentals, and in which the predictions agree closely with experiment. There are few better illustrations of the use of the integral calculus, as well as the principles of ordinary statics, available to the student. A great deal can be done with only elementary mathematics. Properly adapted, the material can be used from the earliest introduction of school science, giving an excellent example of a quantitative science with many possibilities for hands-on experiences.

The definition of a fluid deserves careful consideration. Although time is not a factor in hydrostatics, it enters in the approach to hydrostatic equilibrium. It is usually stated that a fluid is a substance that cannot resist a shearing stress, so that pressures are normal to confining surfaces. Geology has now shown us clearly that there are substances which can resist shearing forces over short time intervals, and appear to be typical solids, but which flow like liquids over long time intervals. Such materials include wax and pitch, ice, and even rock.

A ball of pitch, which can be shattered by a hammer, will spread out and flow in months. Ice, a typical solid, will flow in a period of years, as shown in glaciers, and rock will flow over hundreds of years, as in convection in the mantle of the earth. Shear earthquake waves, with periods of seconds, propagate deep in the earth, though the rock there can flow like a liquid when considered over centuries. The rate of shearing may not be strictly proportional to the stress, but exists even with low stress. Viscosity may be the physical property that varies over the largest numerical range, competing with electrical resistivity. There are several familiar topics in hydrostatics which often appear in expositions of introductory science, and which are also of historical interest that can enliven their presentation.

A study of hydrostatics can also include capillarity, the ideal gas laws, the velocity of sound, and hygrometry. These interesting applications will not be discussed in this article. At a beginning level, it may also be interesting to study the volumes and areas of certain shapes, or at a more advanced level, the forces exerted by heavy liquids on their containers. Hydrostatics is a very concrete science that avoids esoteric concepts and advanced mathematics. It is also much easier to demonstrate than Newtonian mechanics.

Pressure

By a fluid, we have a material in mind like water or air, two very common and important fluids. Water is incompressible, while air is very compressible, but both are fluids. Water has a definite volume; air does not. Water and air have low viscosity; that is, layers of them slide very easily on one another, and they quickly assume their permanent shapes when disturbed by rapid flows. Other fluids, such as molasses, may have high viscosity and take a long time to come to equilibrium, but they are no less fluids. The coefficient of viscosity is the ratio of the shearing force to the velocity gradient. Hydrostatics deals with permanent, time-independent states of fluids, so viscosity does not appear, except as discussed in the Introduction.

A fluid, therefore, is a substance that cannot exert any permanent forces tangential to a boundary. Any force that it exerts on a boundary must be normal to the boundary. Such a force is proportional to the area on which it is exerted, and is called a *pressure*. We can imagine any surface in a fluid as dividing the fluid into parts pressing on each other, as if it were a thin material membrane, and so think of the pressure at any point in the fluid, not just at the boundaries. In order for any small element of the fluid to be in equilibrium, the pressure must be the same in all directions (or the element would move in the direction of least pressure), and if no other forces are acting on the body of the fluid, the pressure must be the same at all neighbouring points.

Therefore, in this case the pressure will be the same throughout the fluid, and the same in any direction at a point (Pascal's Principle). Pressure is expressed in units of force per unit area such as dyne/ cm^2, N/cm^2 (pascal), $pounds/in^2$ (psi) or $pounds/ft^2$ (psf). The axiom that if a certain volume of fluid were somehow made solid, the equilibrium of forces would not be disturbed, is useful in reasoning about forces in fluids.

On earth, fluids are also subject to the force of gravity, which acts vertically downward, and has a magnitude $\gamma = \rho g$ per unit volume, where g is the acceleration of gravity, approximately 981 cm/s^2 or 32.15 ft/s^2, ρ is the *density*, the mass per unit volume, expressed in g/cm^3, kg/m^3, or slug/ft^3, and γ is the *specific weight*, measured in lb/in^3, or lb/ft^3 (pcf). Gravitation is an example of a *body force* that disturbs the equality of pressure in a fluid. The presence of the gravitational body force causes the pressure to increase with depth, according to the equation $dp = \rho g\ dh$, in order to support the water above. We call this relation the *barometric equation*, for when this equation is integrated, we find the variation of pressure with height or depth. If the fluid is incompressible, the equation can be integrated at once, and the pressure as a function of depth h is $p = \rho gh + p_0$. The density of water is about 1 g/cm^3, or its specific weight is 62.4 pcf. We may ask what depth of water gives the normal sea-level atmospheric pressure of 14.7 psi, or 2117 psf. This is simply 2117 / 62.4 = 33.9 ft of water. This is the maximum height to which water can be raised by a suction pump, or, more correctly, can be supported by atmospheric pressure.

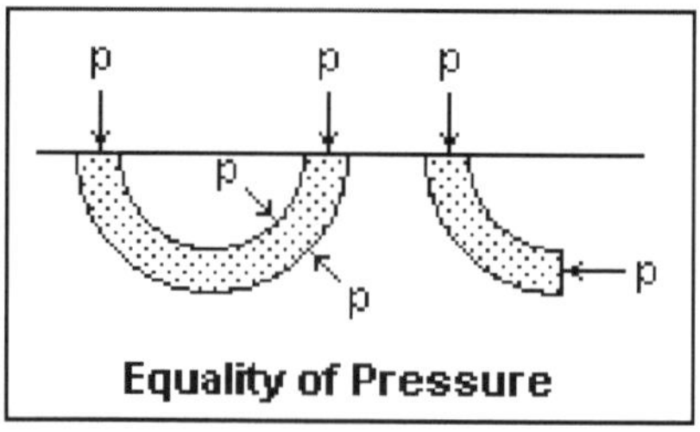

Professor James Thomson (brother of William Thomson, Lord Kelvin) illustrated the equality of pressure by a "curtain-ring" analogy shown in the diagram. A section of the toroid was identified, imagined to be solidified, and its equilibrium was analyzed. The forces exerted on the curved surfaces have no component along the normal to a plane section, so the pressures at any two points of a plane must be equal, since the fluid represented by the curtain ring was in equilibrium. The right-hand part of the diagram illustrates the equality of pressures in orthogonal directions. This can be extended to any direction whatever, so Pascal's Principle is established. This demonstration is similar to the usual one using a triangular prism and considering the forces on the end and lateral faces separately.

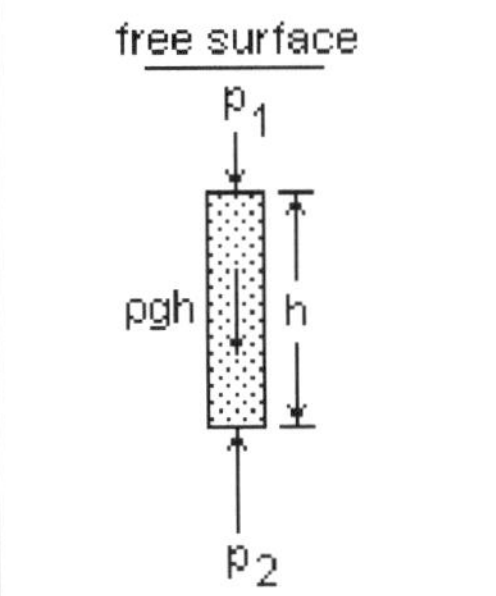

Increase of Pressure with Depth

When gravity acts, the liquid assumes a free surface perpendicular to gravity, which can be proved by Thomson's method. A straight cylinder of unit cross-sectional area (assumed only for ease in the arithmetic) can be used to find the increase of pressure with depth. Indeed, we see that $p_2 = p_1 + \rho gh$. The upper surface of the cylinder can be placed at the free surface if desired. The pressure is now the same in any direction at a point, but is greater at points that lie deeper. From this same figure, it is easy to prove Archimedes's Principle, that the buoyant force is equal to the weight of the displaced fluid, and passes through the centre of mass of this displaced fluid.

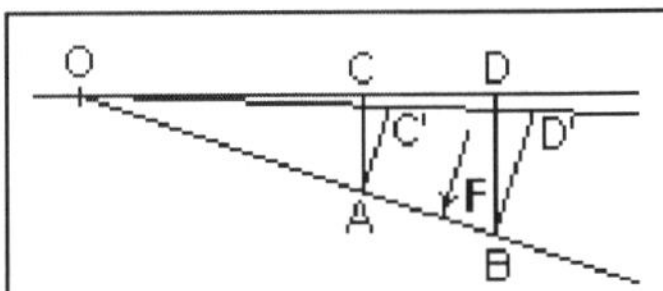

Thrust on a Plane

Ingenious geometric arguments can be used to substitute for easier, but less transparent arguments using calculus. For example, the force on acting on one side of an inclined plane surface whose projection is AB can be found as in the diagram at the right. O is the point at which the prolonged projection intersects the free surface. The line AC' perpendicular to the plane is made equal to the depth AC of point A, and line BD' is similarly drawn equal to BD. The line OD' also passes through C', by proportionality of triangles OAC' and OAD'. Therefore, the thrust F on the plane is the weight of a prism of fluid of cross-section AC'D'B, passing through its centroid normal to plane AB. Note that the thrust is equal to the density times the area times the depth of the centre of the area, but its line of action does not pass through the centre, but below it, at the *centre of thrust*. The same result can be obtained with calculus by summing the pressures and the moments, of course.

Atmospheric Pressure and its Effects

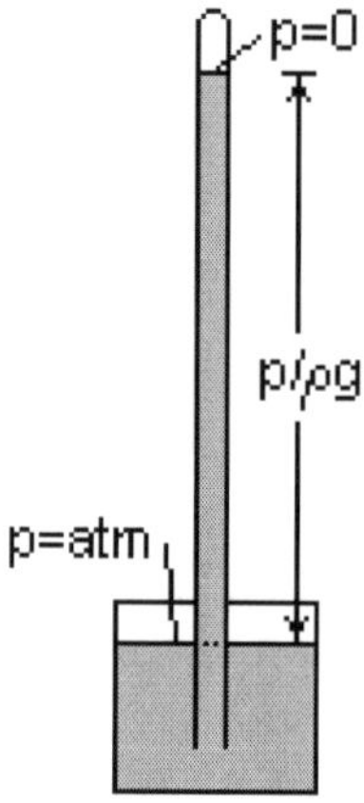

Barometer

Suppose a vertical pipe is stood in a pool of water, and a vacuum pump applied to the upper end. Before we start the pump, the water levels outside and inside the pipe are equal, and the pressures on the surfaces are also equal, and equal to the atmospheric pressure. Now start the pump. When it has sucked all the air out above the water, the pressure on the surface of the water inside the pipe is zero, and the pressure at the level of the water on the outside of the pipe is still the atmospheric pressure.

Of course, there is the vapour pressure of the water to worry about if you want to be precise, but we neglect this complication in making our point. We require a column of water 33.9 ft high inside the pipe, with a vacuum above it, to balance the atmospheric pressure. Now do the same thing with liquid mercury, whose density at 0 °C is 13.5951 times that of water. The height of the column is 2.494 ft, 29.92 in, or 760.0 mm. This definition of the standard atmospheric pressure was established by Regnault in the mid-19th century. In Britain, 30 inHg (inches of mercury) had been used previously.

As a practical matter, it is convenient to measure pressure differences by measuring the height of liquid columns, a practice known as *manometry*. The barometre is a familiar example of this, and atmospheric pressures are traditionally given in terms of the length of a mercury column. To make a barometre, the barometric tube, closed at one end, is filled with mercury and then inverted and placed in a mercury reservoir. Corrections must be made for temperature, because the density of mercury depends on the temperature, and the brass scale expands, for capillarity if the tube is less than about 1 cm in diameter, and even slightly for altitude, since the value of g changes with altitude. The vapor pressure of mercury is only 0.001201 mmHg at 20°C, so a correction from this source is negligible. For the usual case of a mercury column (α = 0.000181792 per °C) and a brass scale (&alpha = 0.0000184 per °C) the temperature correction is -2.74 mm at 760 mm and 20°C. Before reading the barometre scale, the mercury reservoir is raised or lowered until the surface of the mercury just touches a reference point, which is mirrored in the surface so it is easy to determine the proper position.

An *aneroid* barometre uses a partially evacuated chamber of thin metal that expands and contracts according to the external pressure. This movement is communicated to a needle that revolves in a dial. The materials and construction are arranged to give a low temperature coefficient. The instrument must be calibrated before use, and is usually arranged to read directly in elevations. An aneroid barometre is much easier to use in field observations, such as in reconnaisance surveys. In a particular case, it would be read at the start of the day at the base camp, at various points in the vicinity, and then finally at the starting point, to determine the change in pressure with time. The height differences can be calculated from $h = 60{,}360 \log(P/p)\ [1 + (T + t - 64)/986)$ feet, where P and p are in the same units, and T, t are in °F.

An *absolute* pressure is referred to a vacuum, while a *gauge* pressure is referred to the atmospheric pressure at the moment. A negative gauge pressure is a (partial) vacuum. When a vacuum is stated to be so many inches, this means the pressure below the atmospheric pressure of about 30 in.

A vacuum of 25 inches is the same thing as an absolute pressure of 5 inches (of mercury). Pressures are very frequently stated in terms of the height of a fluid. If it is the same fluid whose pressure is being given, it is usually called "head," and the factor connecting the head and the pressure is the weight density ρg. In the English engineer's system, weight density is in pounds per cubic inch or cubic foot. A head of 10 ft is equivalent to a pressure of 624 psf, or 4.33 psi. It can also be considered an *energy availability* of ft-lb per lb. Water with a pressure head of 10 ft can furnish the same energy as an equal amount of water raised by 10 ft. Water flowing in a pipe is subject to *head loss* because of friction.

Take a jar and a basin of water. Fill the jar with water and invert it under the water in the basin. Now raise the jar as far as you can without allowing its mouth to come above the water surface. It is always a little surprising to see that the jar does not empty itself, but the water remains with no visible means of support.

By blowing through a straw, one can put air into the jar, and as much water leaves as air enters. In fact, this is a famous method of collecting insoluble gases in the chemical laboratory, or for supplying hummingbird feeders. It is good to remind oneself of exactly the balance of forces involved.

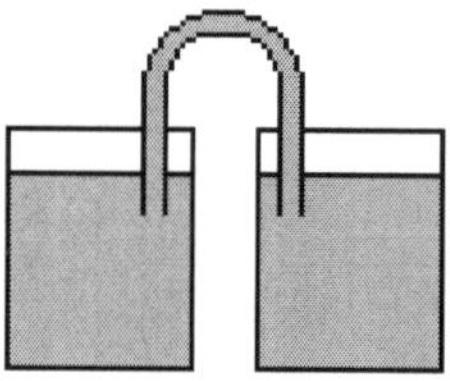

The Siphon

Another application of pressure is the *siphon*. The name is Greek for the tube that was used for drawing wine from a cask. This is a tube filled with fluid connecting two containers of fluid, normally rising higher than the water levels in the two containers, at least to pass over their rims. In the diagram, the two water levels are the same, so there will be no flow. When a siphon goes below the free water levels, it is called an *inverted siphon*.

If the levels in the two basins are not equal, fluid flows from the basin with the higher level into the one with the lower level, until the levels are equal. A siphon can be made by filling the tube, closing the ends, and then putting the ends under the surface on both sides. Alternatively, the tube can be placed in one fluid and filled by sucking on it. When it is full, the other end is put in place. The analysis of the siphon is easy, and should be obvious.

The pressure rises or falls as described by the barometric equation through the siphon tube. There is obviously a maximum height for the siphon which is the same as the limit of the suction pump, about 34 feet. Inverted siphons (which are really not siphons at all) are sometimes used in pipelines to cross valleys.

Differences in elevation are usually too great to use regular siphons to cross hills, so the fluids must be pressurized by pumps so the pressure does not fall to zero at the crests. The Quabbin Aqueduct, which supplies water to Boston, includes pumped siphons.

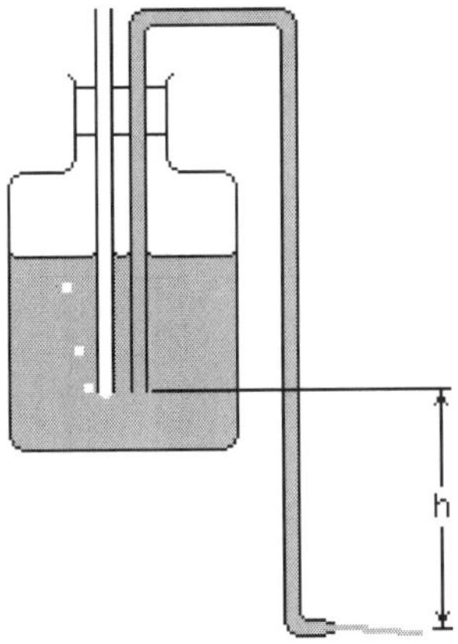

Mariotte's Bottle

As the level in the supply container falls, the pressure difference decreases. In some cases, one would like a source that would provide a constant pressure at the outlet of the siphon. An ingenious way to arrive at this is shown in the figure, Mariotte's Bottle. The plug must seal the air space at the top very well. A partial vacuum is created in the air space by the fall of the water level exactly equal to the pressure difference between the surface and the end of the open tube connecting to the atmosphere. The pressure at this point is, therefore, maintained at atmospheric while water is delivered. The head available at the nozzle as shown is equal to h. This would make a good experiment to verify the relation $V = \sqrt{(2gh)}$ since h and the horizontal distance reached by the jet for a given fall can both be measured easily, or the discharge from an orifice.

The term "siphon" is often used in a different sense. In biology, a siphon is simply a tubular structure. A *thermal siphon* is a means to circulate a liquid by convection. A soda siphon is a source of carbonated water, while siphon coffee (or vacuum coffee) is made in an apparatus where the steam from boiling water pushes hot water up above the coffee and filter, and then the vacuum causes the water to descend again when the heat is removed (invented by Löff in 1830). None of these arrangements is actually a siphon in the physicist's sense. The siphon tube used in irrigation, and perhaps Thomson's siphon recorder of 1858, do use the siphon principle. The occasional spelling "syphon" is not supported by the Greek source.

In some cases, especially in plumbing, siphon action is not desired, especially when it may allow dirty water to mix with clean. In these cases, *vacuum breakers* may be used at high points to prevent this. Siphons work because of atmospheric pressure, and would not operate in a vacuum. In the case of water, pressure reduction would eventually reach the vapor pressure and the water would boil. Mercury, which has a very low vapor pressure, would simply separate leaving a Torricellian vacuum. The siphon would be re-established if the pressure is restored. A liquid column is unstable under a negative pressure.

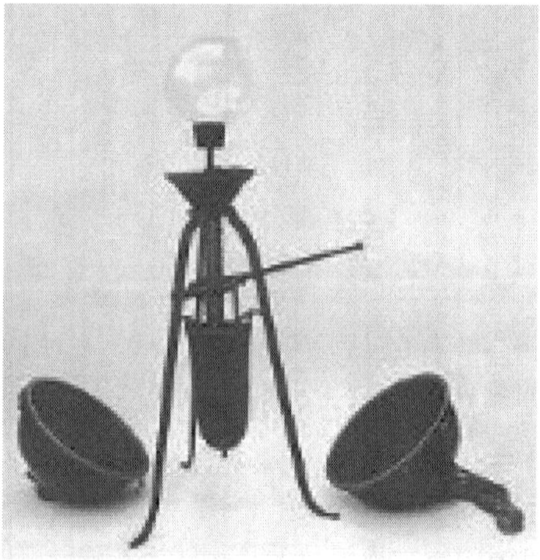

Evangelista Torricelli (1608-1647), Galileo's student and secretary, a member of the Florentine Academy of Experiments, invented the mercury barometre in 1643, and brought the weight of the atmosphere to light. The mercury column was held up by the pressure of the atmosphere, not by *horror vacui* as Aristotle had supposed. Torricelli's early death was a blow to science, but his ideas were furthered by Blaise Pascal (1623-1662).

Pascal had a barometre carried up the 1465 m high Puy de Dôme, an extinct volcano in the Auvergne just west of his home of Clermont-Ferrand in 1648 by Périer, his brother-in-law. Pascal's *experimentum crucis* is one of the triumphs of early modern science. The Puy de Dôme is not the highest peak in the Massif Central—the Puy de Sancy, at 1866 m is, but it was the closest. Clermont is now the centre of the French pneumatics industry.

The remarkable Otto von Guericke (1602-1686), Burgomeister of Magdeburg, Saxony, took up the cause, making the first vacuum pump, which he used in vivid demonstrations of the pressure of the atmosphere to the Imperial Diet at Regensburg in 1654. Famously, he evacuated a sphere consisting of two well-fitting hemispheres about a foot in diameter, and showed that 16 horses, 8 on each side, could not pull them apart. An original vacuum pump and hemispheres from 1663 are shown at the right. He also showed that air had weight, and how much force it did require to separate evacuated hemispheres. Then, in England, Robert Hooke (1635-1703) made a vacuum pump for Robert Boyle (1627-1691). Christian Huygens (1629-1695) became interested in a visit to London in 1661 and had a vacuum pump built for him. By this time, Torricelli's doctrine had triumphed over the Church's support for *horror vacui*. This was one of the first victories for rational physics over the illusions of experience, and is well worth consideration.

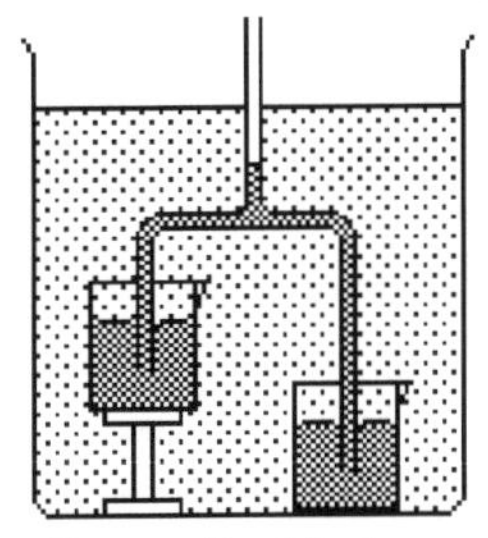

Pascal's Siphon

Pascal demonstrated that the siphon worked by atmospheric pressure, not by *horror vacui*, by means of the apparatus shown at the left. The two beakers of mercury are connected by a three-way tube as shown, with the upper branch open to the atmosphere. As the large container is filled with water, pressure on the free surfaces of the mercury in the beakers pushes mercury into the tubes. When the state shown is reached, the beakers are connected by a mercury column, and the siphon starts, emptying the upper beaker and filling the lower. The mercury has been open to the atmosphere all this time, so if there were any *horror vacui*, it could have flowed in at will to soothe itself.

The mm of mercury is sometimes called a torr after Torricelli, and Pascal also has been honoured by a unit of pressure, a newton per square metre or 10 dyne/cm^2. A cubic centimetre of air weighs 1.293 mg under standard conditions, and a cubic metre 1.293 kg, so air is by no means even approximately weightless, though it seems so. The weight of a sphere of air as small as 10 cm in diameter is 0.68 g, easily measurable with a chemical balance. The pressure of the atmosphere is also considerable, like being 34 ft under water, but we do not notice it. A *bar* is 10^6 dyne/cm^2, very close to a standard atmosphere, which is 1.01325 bar. In meteorolgy, the millibar, mb, is used. 1 mb = 1.333 mmHg = 100 Pa = 1000 dyne/cm^2. A kilogram-force per square centimetre is 981,000 dyne/cm^2, also close to one atmosphere. In Europe, it has been considered approximately 1 atm, as in tire pressures and other engineering applications. As we have seen, in English units the atmosphere is about 14.7 psi, and this figure can be used to find other approximate equivalents. For example, 1 psi = 51.7 mmHg. In Britain, tons per square inch has been used for large pressures. The ton in this case is 2240 lb, not the American short ton. 1 tsi = 2240 psi, 1 tsf = 15.5 psi (about an atmosphere!).

The fluid in question here is air, which is by no means incompressible. As we rise in the atmosphere and the pressure decreases, the air also expands. To see what happens in this case, we can make use of the ideal gas equation of state, $p = \rho RT/M$, and assume that the temperature T is constant. Then the change of pressure in a change of altitude dh is $dp = -\rho g\, dh = -(pM/RT)g dh$, or $dp/p = -(Mg/RT)dh$. This is a little harder to integrate than before, but the result is $\ln p = -Mgh/RT + C$, or $\ln(p/p_0) = -Mgh/RT$, or finally $p = p_0 \exp(-Mgh/RT)$. In an *isothermal* atmosphere, the pressure decreases exponentially. The quantity $H = RT/Mg$ is called the "height of the homogeneous atmosphere" or the *scale height*, and is about 8 km at T = 273K. This quantity gives the rough scale of the decrease of pressure with height. Of course, the real atmosphere is by no means isothermal close to the ground, but cools with height nearly linearly at about 6.5°C/km up to an altitude of about 11 km at middle latitudes, called the *tropopause*. Above this is a region of nearly constant temperature, the stratosphere, and then at some higher level the atmosphere warms again to near its value at the surface. Of course, there are variations from the average values. When the temperature profile with height is known, we can find the pressure by numerical integration quite easily.

The atmospheric pressure is of great importance in meteorology, since it determines the winds, which generally move at right angles to the direction of most rapid change of pressure, that is, *along* the isobars, which are contours of constant pressure. Certain typical weather patterns are associated with relatively high and relatively low pressures, and how they vary with time. The barometric pressure may be given in popular weather forecasts, though few people know what to do with it. If you live at a high altitude, your local weather reporter may report the pressure to be, say, 29.2 inches, but if you have a real barometre, you may well find that it is closer to 25 inches. At an elevation of 1500 m (near Denver, or the top of the Puy de

Dôme), the atmospheric pressure is about 635 mm, and water boils at 95 °C. In fact, altitude is quite a problem in meteorology, since pressures must be measured at a common level to be meaningful. The barometric pressures quoted in the news are *reduced to sea level* by standard formulas, that amount to assuming that there is a column of air from your feet to sea level with a certain temperature distribution, and adding the weight of this column to the actual barometric pressure. This is only an arbitrary 'fix' and leads to some strange conclusions, such as the permanent winter highs above high plateaus that are really imaginary.

The Hydraulic Lever

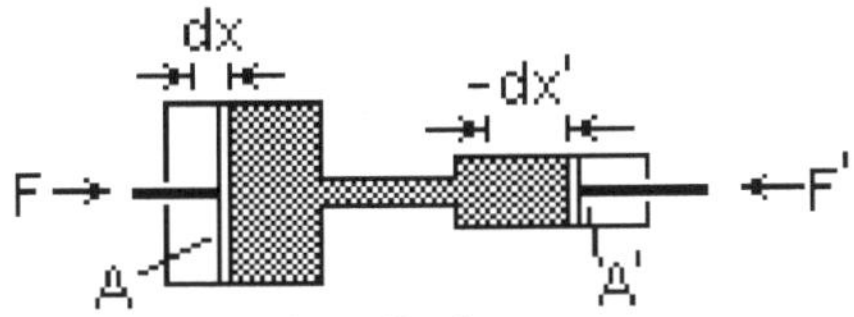

Hydraulic Lever

A cylinder and piston is a chamber of variable volume, a mechanism for transforming pressure to force. If A is the area of the cylinder, and p the pressure of the fluid in it, then F = pA is the force on the piston. If the piston moves outwards a distance dx, then the change in volume is dV = A dx. The work done by the fluid in this displacement is dW = F dx = pA dx = p dV. If the movement is slow enough that inertia and viscosity forces are negligible, then hydrostatics will still be valid. A process for which this is true is called *quasi-static*.

Now consider two cylinders, possibly of different areas A and A', connected with each other and filled with fluid. For simplicity, suppose that there are no gravitational forces. Then the pressure is the same, p, in both cylinders. If the fluid is incompressible, then dV + dV' = 0, so that dW = p dV + p dV' = F dx + F' dx' = 0. This says the work done on one piston is equal to the work done by the other piston: the conservation of energy. The ratio of the forces on the pistons is F' / F = A' / A, the same as the ratio of the areas, and the ratios of the displacements dx' / dx = F / F' = A / A' is in the inverse ratio of the areas. This mechanism is the hydrostatic analogue of the lever, and is the basis of hydraulic activation.

The most famous application of this principle is the Bramah hydraulic press, invented in 1785 by Joseph Bramah (1748-1814), who also invented many other useful machines, including a lock and a toilet. Now, it was not very remarkable to see the possibility of a hydraulic press; what was remarkable was to find a way to seal the large cylinder properly. This was the crucial problem that Bramah solved by his leather seal that was held against the cylinder and the piston by the hydraulic pressure itself.

In the presence of gravity, $p' = p + \rho gh$, where h is the difference in elevation of the two cylinders. Now, $p' dV' = -dV (p + \rho gh) = -p\ dV - (\rho\ dV)gh$, or the net work done in the process is $p' dV' + p\ dV = -dM\ gh$, where dM is the mass of fluid displaced from the lower cylinder to the upper cylinder. Again, energy is conserved if we take into account the potential energy of the fluid. Pumps are seen to fall within the province of hydrostatics if their operation is *quasi-static*, which means that dynamic or inertia forces are negligible.

Pumps

Pumps are used to move or raise fluids. They are not only very useful, but are excellent examples of hydrostatics. Pumps are of two general types, hydrostatic or *positive displacement* pumps, and pumps depending on dynamic forces, such as centrifugal pumps. Here we will only consider positive displacement pumps, which can be understood purely by hydrostatic considerations.

They have a *piston* (or equivalent) moving in a closely-fitting *cylinder*, and forces are exerted on the fluid by motion of the piston. We have already seen an important example of this in the hydraulic lever or hydraulic press, which we have called quasi-static. The simplest pump is the *syringe*, filled by withdrawing the piston and emptied by pressing it back in, as its port is immersed in the fluid or removed from it.

More complicated pumps have *valves* allowing them to work repetitively. These are usually *check valves* that open to allow passage in one direction, and close automatically to prevent reverse flow. There are many kinds of valves, and they are usually the most trouble-prone and complicated part of a pump. The *force pump* has two check valves in the cylinder, one for supply and the other for delivery. The supply valve opens when the cylinder volume increases, the delivery valve when the cylinder volume decreases.

The *lift pump* has a supply valve, and a valve in the piston that allows the liquid to pass around it when the volume of the cylinder is reduced. The delivery in this case is from the upper part of the cylinder which the piston does not enter. *Diaphragm* pumps are force pumps in which the oscillating diaphragm takes the place of the piston. The diaphragm may be moved mechanically, or by the pressure of the fluid on one side of the diaphragm.

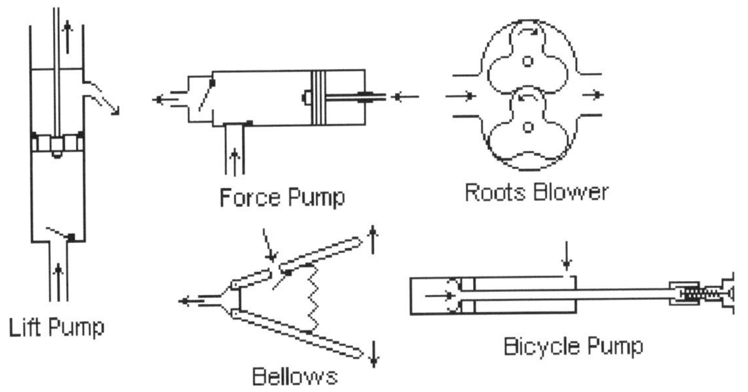

Positive Displacement Pumps

Some positive displacement pumps are shown at the right. The force and lift pumps are typically used for water. The force pump has two valves in the cylinder, while the lift pump has a one valve in the cylinder and one in the piston. The maximum lift, or "suction," is determined by the atmospheric pressure, and either cylinder must be within this height of the free surface.

The force pump, however, can give an arbitrarily large pressure to the discharged fluid, as in the case of a diesel engine injector. A nozzle can be used to convert the pressure to velocity, to produce a jet, as for fire fighting. Fire fighting force pumps usually had two cylinders feeding one receiver alternately. The air space in the receiver helped to make the water pressure uniform.

The three pumps on the right are typically used for air, but would be equally applicable to liquids. The Roots blower has no valves, their place taken by the sliding contact between the rotors and the housing. The Roots blower can either exhaust a receiver or provide air under moderate pressure, in large volumes. The bellows is a very old device, requiring no accurate machining. The single valve is in one or both sides of the expandable chamber. Another valve can be placed at the nozzle if required. The valve can be a piece of soft leather held close to holes in the chamber.

The bicycle pump uses the valve on the valve stem of the tire or inner tube to hold pressure in the tire. The piston, which is attached to the discharge tube, has a flexible seal that seals when the cylinder is moved to compress the air, but allows air to pass when the movement is reversed. Diaphragm and vane pumps are not shown, but they act the same way by varying the volume of a chamber, and directing the flow with check valves.

Pumps were applied to the dewatering of mines, a very necessary process as mines became deeper. Newcomen's atmospheric engine was invented to supply the power for pumping. The first engine may have been erected in Cornwall in 1710, but the Dudley Castle engine of 1712 is much better known and thoroughly documented. The first pumps used in Cornwall were called *bucket* pumps, which we recognise as lift pumps, with the pistons somewhat miscalled buckets. They pumped on the up-stroke, when a clack in the bottom of the pipe opened and allowed water to enter beneath the piston. At the same time, the piston lifted the column of water above it, which could be of any length. The piston could only "suck" water 33 ft, or 28 ft more practically, of course, but this occurred at the bottom of the shaft, so this was only a limit on the piston stroke. On the down stroke, a clack in the bucket opened, allowing it to sink through the water to the bottom, where it would be ready to make another lift.

More satisfactory were the *plunger* pumps, also placed at the bottom of the shaft. A plunger displaced volume in a chamber, forcing the water in it through a check valve up the shaft, when it descended. When it rose, water entered the pump chamber through a clack, as in the bucket pump. Only the top of the plunger had to be packed; it was not necessary that it fit the cylinder accurately. In this case, the engine at the surface lifted the heavy pump rods on the up-stroke. When the atmospheric engine piston returned, the heavy timber pump rods did the actual pumping, borne down by their weight.

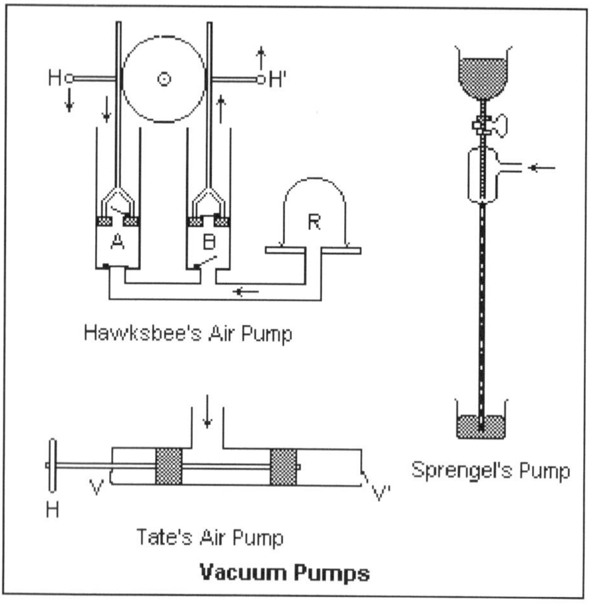

Vacuum Pumps

A special application for pumps is to produce a vacuum by exhausting a container, called the *receiver*. Hawksbee's dual cylinder pump, designed in the 18th century, is the final form of the air pump invented by Guericke by 1654. A good pump could probably reach about 5-10 mmHg, the limit set by the valves. The cooperation of the cylinders made the pump much easier to work when the pressure was low. In the diagram, piston A is descending, helped by the partial vacuum remaining below it, while piston B is rising, filling with the low-pressure air from the receiver.

The bell-jar receiver, invented by Huygens, is shown; previously, a cumbersome globe was the usual receiver. Tate's air pump is a 19th century pump that would be used for simple vacuum demonstrations and for utility purposes in the lab. It has no valves on the low-pressure side, just exhaust valves V, V', so it could probably reach about 1 mmHg. It is operated by pushing and pulling the handle H. At the present day, motor-driven rotary-seal pumps sealed by running in oil are used for the same purpose. At the right is Sprengel's pump, with the valves replaced by drops of mercury. Small amounts of gas are trapped at the top of the fall tube as the mercury drops, and moves slowly down the fall tube as mercury is steadily added, coming out at the bottom carrying the air with it.

The length of the fall tube must be greater than the barometric height, of course. Theoretically, a vacuum of about 1 μm can be obtained with a Sprengel pump, but it is very slow and can only evacuate small volumes. Later, Langmuir's mercury diffusion pump, which was much faster, replaced Sprengel pumps, and led to oil diffusion pumps that can reach very high vacua.

The *column of water* or *hydrostatic* engine is the inverse of the force pump, used to turn a large head (pressure) of water into rotary motion. It looks like a steam engine, with valves operated by valve gear, but of course is not a heat engine and can be of high efficiency. However, it is not of as high efficiency as a turbine, and is much more complicated, but has the advantage that it can be operated at variable speeds, as for lifting. A few very impressive column of water engines were made in the 19th century, but they were never popular and remained rare. Richard Trevithick, famous for high pressure steam engines, also built hydrostatic engines in Cornwall.

The photograph at the right shows a column-of-water engine built by George von Reichenbach, and placed in service in 1917. This engine was exhibited in the Deutsches Museum in München as late as 1977. It was used to pump brine for the Bavarian state salt industry. A search of the museum website did not reveal any evidence of it, but a good drawing of another brine pump, with four cylinders and driven by a water wheel, also built by von Reichenbach was found. This machine, a *Solehebemaschine* ("brine-lifting machine"), entered service in 1821. It had two pressure-operated poppet valves for each cylinder. These engines are brass to resist corrosion by the salt water. Water pressure engines must be designed taking into account the incompressibility of water, so both valves must not close at the same time, and abrupt changes of rate of flow must not be made. Air chambers can be used to eliminate shocks.

Georg von Reichenbach (1771-1826) is much better known as an optical designer than as a mechanical engineer. He was associated with Joseph Fraunhofer, and they died within days of each other in 1826. He was of an aristocratic family, and was *Salinenrat*, or manager of the state salt works, in southeastern Bavaria, which was centred on the town of Reichenhall, now Bad Reichenhall, near Salzburg. The name derives from "rich in salt." This famous salt region had salt springs flowing nearly saturated brine, at 24% to 26% (saturated is 27%) salt, that from ancient times had been evaporated over wood fires. A brine pipeline to Traunstein was constructed in 1617-1619, since wood fuel for evaporating the brine was exhausted in Reichenhall.

The pipeline was further extended to Rosenheim, where there was turf as well as wood, in 1818-10. Von Reichenbach is said to have built this pipeline, for which he designed a water-wheel-driven, four-barrel pump. Maximilian I, King of Bavaria, commissioned von Reichenbach to bring brine from Berchtesgaden, elevation 530 m, to Reichenhall, elevation 470 m, over a summit 943 m high. The pump shown in the photograph pumped brine over this line, entering service in 1816. Fresh water was also allowed to flow down to the salt beds, and the brine was then pumped to the surface. This was a much easier way to mine salt than underground mining. The salt industry of Bad Reichenhall still operates, but it is now Japanese-owned.

Forces on Submerged Surfaces

Suppose we want to know the force exerted on a vertical surface of any shape with water on one side, assuming gravity to act, and the pressure on the surface of the water zero. We have already solved this problem by a geometrical argument, but now we apply calculus, which is easier but not as illuminating. The force on a small area dA a distance x below the surface of the water is $dF = p\ dA = \rho gx\ dA$, and the moment of this force about a point on the surface is $dM = px\ dA = \rho gx^2\ dA$. By integration, we can find the total force F, and the depth at which it acts, $c = M / F$. If the surface is not symmetrical, the position of the total force in the transverse direction can be obtained from the integral of $dM' = \rho gxy\ dA$, the moment about some vertical line in the plane of the surface. If there happens to be a pressure on the free surface of the water, then the forces due to this pressure can be evaluated separately and added to this result. We must add a force equal to the area of the surface times the additional pressure, and a moment e qual to the product of this force and the distance to the centroid of the surface.

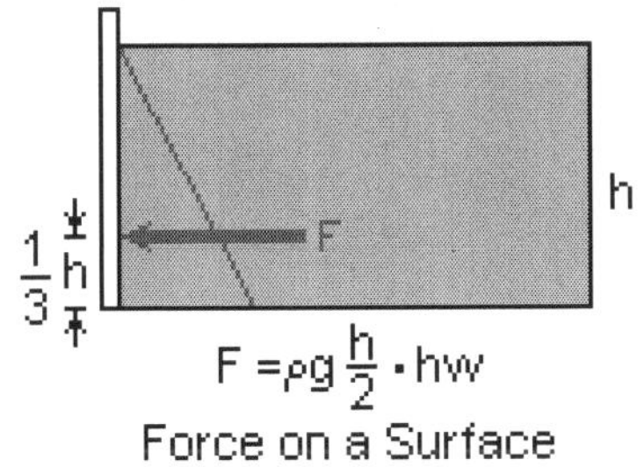

$F = \rho g \frac{h}{2} \cdot hw$

Force on a Surface

The simplest case is a rectangular gate of width w, and height h, whose top is a distance H below the surface of the water. In this case, the integrations are very easy, and $F = \rho gw[(h + H)^2 - h^2]/2 =$

$\rho gH(H + 2h)/2 = \rho g(h + H/2)Hw$. The total force on the gate is equal to its area times the pressure at its centre. $M = \rho gw[(h + H)^3 - h^3]/3 = \rho g(H^2/3 + Hh + h^2)Hw$, so that $c = (H^2/3 + Hh + h^2)/(h + H/2)$. In the simple case of $h = 0$, $c = 2H/3$, or two-thirds of the way from the top to the bottom of the gate. If we take the atmospheric pressure to act not only on the surface of the water, but also the dry side of the gate, there is no change to this result. This is the reason atmospheric pressure often seems to have been neglected in solving subh problems.

Hydrostatic Paradox

Consider a curious rectangular tank, with one side vertical but the opposite side inclined inwards or outwards. The horizontal forces exerted by the water on the two sides must be equal and opposite, or the tank would scoot off. If the side is inclined outwards, then there must be a downwards vertical force equal to the weight of the water above it, and passing through the centroid of this water. If the side is inclined inwards, there must be an upwards vertical force equal to the weight of the 'missing' water above it. In both cases, the result is demanded by ordinary statics. What we have here has been called the 'hydrostatic paradox.' It was conceived by the celebrated Flemish engineer Simon Stevin (1548-1620) of Brugge, the first modern scientist to investigate the statics of fluids and solids. Consider three tanks with bottoms of equal sizes and equal heights, filled with water. The pressures at the bottoms are equal, so the vertical force on the bottom of each tank is the same. But suppose that one tank has vertical sides, one has sides inclined inward, and third sides inclined outwards. The tanks do not contain the same weight of water, yet the forces on their bottoms are equal! I am sure that you can spot the resolution of this paradox.

Sometimes the forces are required on curved surfaces. The vertical and horizontal components can be found by considering the equilibrium of volumes with a plane surface equal to the projected area of the curved surface in that direction. The general result is usually a force plus a couple, since the horizontal and vertical forces are not necessarily in the same plane. Simple surfaces, such as cylinders, spheres and cones, may often be easy to solve. In general, however, it is necessary to sum the forces and moments numerically on each element of area, and only in simple cases can this be done analytically.

If a volume of fluid is accelerated uniformly, the acceleration can be added to the acceleration of gravity. A free surface now becomes perpendicular to the total acceleration, and the pressure is proportional to the distance from this surface. The same can be done for a rotating fluid, where the centrifugal acceleration is the important quantity. The earth's atmosphere is an example. When air moves relative to the rotating system, the Coriolis force must also be taken into account. However, these are dynamic effects and are not strictly a part of hydrostatics.

Buoyancy

Archimedes, so the legend runs, was asked to determine if the goldsmith who made a golden crown for Hieron, Tyrant of Syracuse, had substituted cheaper metals for gold. The story is told by Vitruvius. A substitution could not be detected by simply weighing the crown, since it was craftily made to the same weight as the gold supplied for its construction. Archimedes realised that finding the density of the crown, that is, the weight per unit volume, would give the answer. The weight was known, of course, and Archimedes cunningly measured its volume by the amount of water that ran off when it was immersed in a vessel filled to the brim. By comparing the results for the crown, and for pure gold, it was found that the crown displaced more water than an equal weight of gold, and had, therefore, been adulterated.

This story, typical of the charming way science was made more interesting in classical times, may or may not actually have taken place, but whether it did or not, Archimedes taught that a body immersed in a fluid lost apparent weight equal to the weight of the fluid displaced, called *Archimedes' Principle.* Specific gravity, the ratio of the density of a substance to the density of water, can be determined by weighing the body in air, and then in water. The specific gravity is the weight in air divided by the loss in weight when immersed. This avoids the difficult determination of the exact volume of the sample.

To see how buoyancy works, consider a submerged brick, of height h, width w and length l. The difference in pressure on top and bottom of the brick is ρgh, so the difference in total force on top and bottom of the brick is simply $(\rho gh)(wl) = \rho gV$, where V is the volume of the brick. The forces on the sides have no vertical components, so they do not matter. The net upward force is the weight of a volume V of the fluid of density ρ. Any body can be considered made up of brick shapes, as small as desired, so the result applies in general. This is just the integral calculus in action, or the application of Professor Thomson's analogy.

Consider a man in a rowboat on a lake, with a large rock in the boat. He throws the rock into the water. What is the effect on the water level of the lake? Suppose you make a drink of ice water with ice cubes floating in it. What happens to the water level in the glass when the ice has melted?

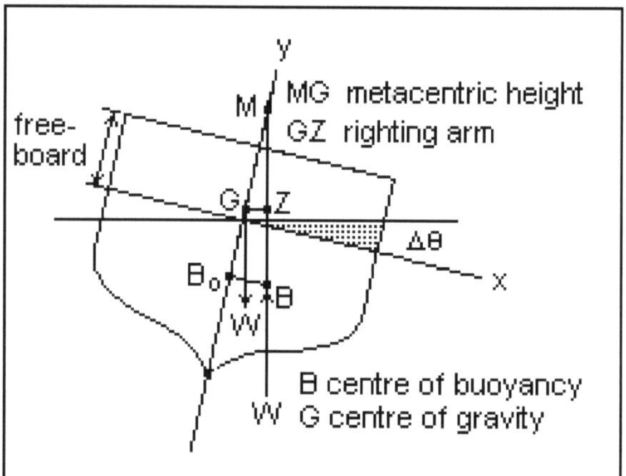

Ship Stability

The force exerted by the water on the bottom of a boat acts through the centre of gravity B of the displaced volume, or centre of buoyancy, while the force exerted by gravity on the boat acts through its own centre of gravity G. This looks bad for the boat, since the boat's c.g. will naturally be higher than the c.g. of the displaced water, so the boat will tend to capsize.

Well, a board floats, and can tell us why. Should the board start to rotate to one side, or *heel*, the displaced volume immediately moves to that side, and the buoyant force tends to correct the rotation. A floating body will be stable provided the line of action of the buoyant force passes through a point M above the c.g. of the body, called the *metacentre*, so that there is a restoring *couple* when the boat heels. A ship with an improperly designed hull will not float. It is not as easy to make boats as it might appear.

Let B_0 be the centre of buoyancy with the ship upright; that is, it is the centre of gravity of the volume V of the displaced water. $\gamma V = W$, the weight of the ship. If the ship heels by an angle $\Delta\theta$, a wedge-shaped volume of water is added on the right, and an equal volume is removed on the left, so that V remains constant. The centre of buoyancy is then moved to the right to point B. We can find the x-coordinate of B by taking moments of the volumes about the y-axis. Therefore, V (B_0B) = V(0) + moment of the shaded volume - moment of the equal compensating volume. If dA is an element of area in the y=0 plane, then the volume element is xΔθdA (this automatically makes the volume to the left of x=0 negative), and the moment is this times x. Note that contributions from $x>0$ and $x<0$ are both positive, as they should be. Now, +”x^2dA is just the moment of inertia of the water-level area of the ship, I. Therefore, V(B_0B) = IΔθ. Now, (B_0B)/Δθ = (MB_0), since for small Δθ the tangent is equal to the angle. Finally, then, (MG) = (I/V) - (B_0G).

The moment tending to restore the ship to upright is W times the *righting arm* GZ = MG x Δθ. Therefore, the ship tends to roll with a certain period. A small GM means a small restoring torque, and so a long roll period. A ship with a small GM is said to be *tender*, which is desirable for passenger ships and for gun platforms (warships). A passenger ship may have a roll period of 28s or so, while a cargo ship may have a period of 13-15s. A ship with a large GM and a short roll period is called *stiff*. Metacentric heights are typically 1 to 2 metres.

The combination of a small GM and a small freeboard was originally considered desirable for a warship, since it made a stable gun platform and presented a minimum area that had to be armoured. HMS Captain, an early turret ironclad launced in 1869, was such a ship. The ship capsized off Finisterre in 1870 in a gale when the topsails were not taken in promptly enough and the ship heeled beyond its 14° maximum. HMS Sultan, a broadside ironclad launched in 1870, had metacentric height of only 3 feet for stability, but proved unsafe for Atlantic service.

The *free surface effect* can greatly reduce the stability of a ship. For example, if the hull has taken water, when the ship heels this weight moves to the low side and counters the buoyancy that should give the ship stability. Longitudinal baffles reduce the effect (division into thirds reduces the effect by a factor of 9), and are absolutely necessary for ships like tankers. In 2006, imprudent shifting of ballast water caused MV Cougar Ace, with its cargo of Mazdas, to list 80°. The ship was eventually righted, however, since it did not take water. A *list*, incidentally, is a permanent heel.

Longitudinal stability against pitching is analyzed similarly.

Archimedes's Principle can also be applied to balloons. The Montgolfier brothers' hot air balloon with a paper envelope ascended first in 1783 (the brothers got Pilâtre de Rozier and Chevalier d'Arlandes to go up in it). Such "fire balloons" were then replaced with hydrogen-filled balloons, and then with balloons filled with coal gas, which was easier to obtain and did not diffuse through the envelope quite as rapidly. Methane would be a good filler, with a density 0.55 that of air. Slack balloons, like most large ones, can be contrasted with taut balloons with an elastic envelope, such as weather balloons. Slack balloons will not be filled full on the ground, and will plump up at altitude. Balloons are naturally stable, since the centre of buoyancy is above the centre of gravity in all practical balloons. Submarines are yet another application of buoyancy, with their own characteristic problems.

Small neoprene or natural rubber balloons have been used for meteorological observations, with hydrogen filling. A 10g *ceiling* balloon was about 17" in diameter when inflated to have a free lift of 40g. It ascended 480ft the first minute, 670ft in a minute and a half, and 360ft per minute afterwards, to find cloud ceilings by timing, up to 2500ft, when it subtended about 2' of arc, easily seen in binoculars. Large *sounding* balloons were used to lift a radiosonde and a parachute for its recovery. An AN/AMT-2 radiosonde of the 1950's weighed 1500g, the paper parachute 100g, and the balloon 350g. The balloon was inflated to give 800g free lift, so it would rise 700-800 ft/min to an altitude of about 50,000 ft (15 km) before it burst. This balloon was about 6 ft in diameter when inflated at the surface, 3 ft in diameter before inflation. The information was returned by radio telemetry, so the balloon did not have to be followed optically. Of intermediate size was the *pilot* balloon, which was followed with a theodolite to determine wind directions and speeds. At night, a pilot balloon could carry a light for ceiling determinations.

The greatest problem with using hydrogen for lift is that it diffuses rapidly through many substances. Weather balloons had to be launched promptly after filling, or the desired free lift would not be obtained. Helium is a little better in this respect, but it also diffuses rapidly. The lift obtained with helium is almost the same as with hydrogen (density 4 compared to 2, where air is 28.97). However, helium is exceedingly rare, and only its unusual occurrence in natural gas from Kansas makes it available. Great care must be taken when filling balloons with hydrogen to avoid sparks and the accumulation of hydrogen in air, since hydrogen is exceedingly flammable and explosive over a wide range of concentrations. Helium has the great advantage that it is not inflammable.

The hydrogen for filling weather balloons came from compressed gas in cylinders, from the reaction of granulated aluminium with sodium hydroxide and water, or from the reaction of calcium hydroxide with water. The chemical reactions are $2Al + 2NaOH + 2H_2O \rightarrow 2NaAlO_2 + 3H_2$, or $CaH_2 + 2H_2O \rightarrow Ca(OH)_2 + 2H_2$. In the first, silicon or zinc could be used instead of aluminium, and in the second, any similar metal hydride. Both are rather expensive sources of hydrogen, but very convenient when only small amounts are required. Most hydrogen is made from the catalytic decomposition of hydrocarbons, or the reaction of hot coke with steam. Electrolysis of water is an expensive source, since more energy is used than is recovered with the hydrogen. Any enthusiasm for a "hydrogen economy" should be tempered by the fact that there are no hydrogen wells, and all the hydrogen must be made with an input of energy usually greater than that available from the hydrogen, and often with the appearance of carbon. Although about 60,000 Btu/lb is available from hydrogen, compared to 20,000 Btu/lb from gasoline, hydrogen compressed to 1000 psi requires 140 times as much volume for the same weight as gasoline. For the energy content of a 13-gallon gasoline tank, a 600-gallon hydrogen tank would be required. The critical temperature of hydrogen is 32K, so liquid storage is out of the question for general use.

Measurement of Specific Gravity

The *specific gravity* of a material is the ratio of the mass (or weight) of a certain sample of it to the mass (or weight) of an equal volume of water, the conventional reference material. In the metric system, the density of water is 1 g/cc, which makes the specific gravity numerically equal to the density. Strictly speaking, density has the dimensions g/cc, while specific gravity is a dimensionless ratio. However, in casual speech the two are often confounded. In English units, however, density, perhaps in lb/cuft or pcf, is numerically different from the specific gravity, since the weight of water is 62.5 lb/cuft.

Things are complicated by the variation of the density of water with temperature, and also by the confusion that gave us the distinction between cc and ml. The milliliter is the volume of 1.0 g of water at 4°C, by definition. The actual volume of 1.0 g of water at 4°C is 0.999973 cm^3 by measurement. Since most densities are not known, or needed, to more than three significant figures, it is clear that this difference is of no practical importance, and the ml can be taken equal to the cc. The density of water at 0°C is 0.99987 g/ml, at 20° 0.99823, and at 100°C 0.95838. The temperature dependence of the density may have to be taken into consideration in accurate work. Mercury, while we are at it, has a density 13.5955 at 0°C, and 13.5461 at 20°C.

The basic idea in finding specific gravity is to weigh a sample in air, and then immersed in water. Then the specific gravity is W/(W - W'), if W is the weight in air, and W' the weight immersed. The denominator is just the buoyant force, the weight of a volume of water equal to the volume of the sample. This can be carried out with an ordinary balance, but special balances, such as the Jolly balance, have been created specifically for this application. Adding an extra weight to the sample allows measurement of specific gravities less than 1.

A *pycnometre* is a flask with a close-fitting ground glass stopper with a fine hole through it, so a given volume can be accurately obtained. The name comes from the Greek puknos, a word meaning "density." If the flask is weighed empty, full of water, and full of a liquid whose specific gravity is desired, the specific gravity of the liquid can easily be calculated. A sample in the form of a powder, to which the usual method of weighing cannot be used, can be put into the pycnometre. The weight of the powder and the weight of the displaced water can be determined, and from them the specific gravity of the powder.

The specific gravity of a liquid can be found with a collection of small weighted, hollow spheres that will just float in certain specific gravities. The closest spheres that will just float and just sink put limits on the specific gravity of the liquid. This method was once used in Scotland to determine the amount of alcohol in distilled liquors. Since the density of a liquid decreases as the temperature increases, the spheres that float are an indication of the temperature of the liquid. Galileo's thermometre worked this way.

A better instrument is the *hydrometre*, which consists of a weighted float and a calibrated stem that protrudes from the liquid when the float is entirely immersed. A higher specific gravity will result in a greater length of the stem above the surface, while a lower specific gravity will cause the hydrometre to float lower. The small cross-sectional area of the stem makes the instrument very sensitive. Of course, it must be calibrated against standards. In most cases, the graduations ("degrees") are arbitrary and reference is made to a table to determine the specific gravities. Hydrometres are used to determine the specific gravity of lead-acid battery electrolyte, and the concentration of antifreeze compounds in engine coolants, as well as the alcohol content of whiskey.

Fluid Dynamics

In physics, fluid dynamics is a subdiscipline of fluid mechanics that deals with fluid flow—the natural science of fluids (liquids and gases) in motion. It has several subdisciplines itself, including aerodynamics (the study of air and other gases in motion) and hydrodynamics (the study of liquids in motion). Fluid dynamics has a wide range of applications, including calculating forces and moments on aircraft, determining the mass flow rate of petroleum through pipelines, predicting weather patterns, understanding nebulae in interstellar space and reportedly modelling fission weapon detonation. Some of its principles are even used in traffic engineering, where traffic is treated as a continuous fluid.

Fluid dynamics offers a systematic structure—which underlies these practical disciplines—that embraces empirical and semi-empirical laws derived from flow measurement and used to solve practical problems. The solution to a fluid dynamics problem typically involves calculating various properties of the fluid, such as velocity, pressure, density, and temperature, as functions of space and time. Before the twentieth century, *hydrodynamics* was synonymous with fluid dynamics. This is still reflected in names of some fluid dynamics topics, like magnetohydrodynamics and hydrodynamic stability, both of which can also be applied to gases.

Equations of Fluid Dynamics

The foundational axioms of fluid dynamics are the conservation laws, specifically, conservation of mass, conservation of linear momentum (also known as Newton's Second Law of Motion), and conservation of energy (also known as First Law of Thermodynamics). These are based on classical mechanics and are modified in quantum mechanics and general relativity. They are expressed using the Reynolds Transport Theorem.

In addition to the above, fluids are assumed to obey the *continuum assumption*. Fluids are composed of molecules that collide with one another and solid objects. However, the continuum assumption considers fluids to be continuous, rather than discrete. Consequently, properties such as density, pressure, temperature, and velocity are taken to be well-defined at infinitesimally small points, and are assumed to vary continuously from one point to another. The fact that the fluid is made up of discrete molecules is ignored.

For fluids which are sufficiently dense to be a continuum, do not contain ionized species, and have velocities small in relation to the speed of light, the momentum equations for Newtonian fluids are the Navier-Stokes equations, which is a non-linear set of differential equations that describes the flow of a fluid whose stress depends linearly on velocity gradients and pressure. The unsimplified equations do not have a general closed-form solution, so they are primarily of use in Computational Fluid Dynamics. The equations can be simplified in a number of ways, all of which make them easier to solve. Some of them allow appropriate fluid dynamics problems to be solved in closed form.

In addition to the mass, momentum, and energy conservation equations, a thermodynamical equation of state giving the pressure as a function of other thermodynamic variables for the fluid is required to completely specify the problem. An example of this would be the perfect gas equation of state:

$$p = \frac{\rho R_u T}{M}$$

where p is pressure, ρ is density, R_u is the gas constant, M is the molar mass and T is temperature.

Conservation Laws

Three conservation laws are used to solve fluid dynamics problems, and may be written in integral or differential form. Mathematical formulations of these conservation laws may be interpreted by considering the concept of a *control volume*. A control volume is a specified volume in space through which air can flow in and out. Integral formulations of the conservation laws consider the change in mass, momentum, or energy within the control volume. Differential formulations of the conservation laws apply Stokes' theorem to yield an expression which may be interpreted as the integral form of the law applied to an infinitesimal volume at a point within the flow.

- Mass continuity (conservation of mass): The rate of change of fluid mass inside a control volume must be equal to the net rate of fluid flow into the volume. Physically, this statement requires that mass is neither created nor destroyed in the control volume, and can be translated into the integral form of the continuity equation:

$$\frac{\partial}{\partial t} \oiiint_V \rho dV = \text{“} \oiint_S \rho \mathrm{u} \cdot dS$$

Above, ρ is the fluid density, u is a velocity vector, and t is time. The left-hand side of the above expression contains a triple integral over the control volume, whereas the right-hand side contains a double integral over the surface of the control volume. The differential form of the continuity equation is:

$$\frac{\partial \rho}{\partial t} + \nabla \cdot (\rho \mathrm{u}) = 0$$

- Conservation of Momentum: This equation applies Newton's second law of motion to the control volume, requiring that any change in momentum of the air within a control volume be due to the net flow of air into the volume and the action of external forces on the air within the volume. In the integral formulation of this equation, body forces here are represented by f_{body}, the body force per unit mass. Surface forces, such as viscous forces, are represented by F_{surf} , the net force due to stresses on the control volume surface.

$$\frac{\partial}{\partial t} \oiiint_V \rho \mathrm{u} dV = -\oiint_S (\rho \mathrm{u} \cdot d\mathrm{S})\mathrm{u} - \oiint_S p d\mathrm{S} + \oiiint_V \rho \mathrm{f}_{body} dV + \mathrm{F}_{surf}$$

The differential form of the momentum conservation equation is as follows. Here, both surface and body forces are accounted for in one total force, *F*. For example, *F* may be expanded into an expression for the frictional and gravitational forces acting on an internal flow.

$$\frac{D\mathrm{u}}{Dt} = \mathrm{F} - \frac{\nabla p}{\rho}$$

In aerodynamics, air is assumed to be a Newtonian fluid, which posits a linear relationship between the shear stress (due to internal friction forces) and the rate of strain of the fluid. The equation above is a vector equation: in a three dimensional flow, it can be expressed as three scalar equations. The conservation of momentum equations for the compressible, viscous flow case are called the Navier-Stokes equations.

- Conservation of Energy: Although energy can be converted from one form to another, the total energy in a given closed system remains constant.

$$\rho \frac{Dh}{Dt} = \frac{Dp}{Dt} + \nabla \cdot (k \nabla T) + \Phi$$

Above, h is enthalpy, k is the thermal conductivity of the fluid, T is temperature, and Φ is the viscous dissipation function. The viscous dissipation function governs the rate at which mechanical energy of the flow is converted to heat. The second law of thermodynamics requires that the dissipation term is always positive: viscosity cannot create energy within the control volume. The expression on the left side is a material derivative.

Compressible vs Incompressible Flow

All fluids are compressible to some extent, that is, changes in pressure or temperature will result in changes in density. However, in many situations the changes in pressure and temperature are sufficiently small that the changes in density are negligible. In this case the flow can be modelled as an incompressible flow. Otherwise the more general compressible flow equations must be used. Mathematically, incompressibility is expressed by saying that the density ρ of a fluid parcel does not change as it moves in the flow field, i.e.,

$$\frac{D\rho}{Dt} = 0,$$

where *D/Dt* is the substantial derivative, which is the sum of local and convective derivatives. This additional constraint simplifies the governing equations, especially in the case when the fluid has a uniform density.

For flow of gases, to determine whether to use compressible or incompressible fluid dynamics, the Mach number of the flow is to be evaluated. As a rough guide, compressible effects can be ignored at Mach numbers below approximately 0.3. For liquids, whether the incompressible assumption is valid depends on the fluid properties (specifically the critical pressure and temperature of the fluid) and the flow conditions (how close to the critical pressure the actual flow pressure becomes). Acoustic problems always require allowing compressibility, since sound waves are compression waves involving changes in pressure and density of the medium through which they propagate.

Viscous vs Inviscid Flow

Viscous problems are those in which fluid friction has significant effects on the fluid motion. The Reynolds number, which is a ratio between inertial and viscous forces, can be used to evaluate whether viscous or inviscid equations are appropriate to the problem. Stokes flow is flow at very low Reynolds numbers, $Re \ll 1$, such that inertial forces can be neglected compared to viscous forces. On the contrary, high Reynolds numbers indicate that the inertial forces are more significant than the viscous (friction) forces. Therefore, we may assume the flow to be an inviscid flow, an approximation in which we neglect viscosity completely, compared to inertial terms.

This idea can work fairly well when the Reynolds number is high. However, certain problems such as those involving solid boundaries, may require that the viscosity be included. Viscosity often cannot be neglected near solid boundaries because the no-slip condition can generate a thin region of large strain rate (known as Boundary layer) which enhances the effect of even a small amount of viscosity, and thus generating vorticity. Therefore, to calculate net forces on bodies (such as wings) we should use viscous flow equations. As illustrated by d'Alembert's paradox, a body in an inviscid fluid will experience no drag force. The standard equations of inviscid flow are the Euler equations. Another often used model, especially in computational fluid dynamics, is to use the Euler equations away from the body and the boundary layer equations, which incorporates viscosity, in a region close to the body.

The Euler equations can be integrated along a streamline to get Bernoulli's equation. When the flow is everywhere irrotational and inviscid, Bernoulli's equation can be used throughout the flow field. Such flows are called potential flows.

Steady vs Unsteady Flow

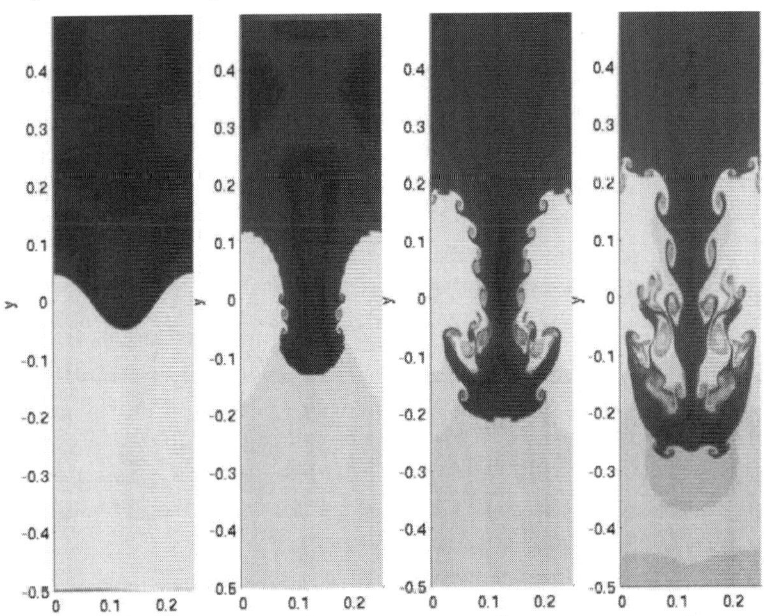

Figure: *Hydrodynamics simulation of the Rayleigh–Taylor instability*

When all the time derivatives of a flow field vanish, the flow is considered to be a steady flow. Steady-state flow refers to the condition where the fluid properties at a point in the system do not change over time. Otherwise, flow is called unsteady. Whether a particular flow is steady or unsteady, can depend on the chosen frame of reference. For instance, laminar flow over a sphere is steady in the frame of reference that is stationary with respect to the sphere. In a frame of reference that is stationary with respect to a background flow, the flow is unsteady.

Turbulent flows are unsteady by definition. A turbulent flow can, however, be statistically stationary. According to Pope:

The random field $U(x,t)$ is statistically stationary if all statistics are invariant under a shift in time.

This roughly means that all statistical properties are constant in time. Often, the mean field is the object of interest, and this is constant too in a statistically stationary flow.

Steady flows are often more tractable than otherwise similar unsteady flows. The governing equations of a steady problem have one dimension fewer (time) than the governing equations of the same problem without taking advantage of the steadiness of the flow field.

Laminar vs Turbulent Flow

Turbulence is flow characterized by recirculation, eddies, and apparent randomness. Flow in which turbulence is not exhibited is called laminar. It should be noted, however, that the presence of eddies or recirculation alone does not necessarily indicate turbulent flow—these phenomena may be present in laminar flow as well. Mathematically, turbulent flow is often represented via a Reynolds decomposition, in which the flow is broken down into the sum of an average component and a perturbation component.

It is believed that turbulent flows can be described well through the use of the Navier–Stokes equations. Direct numerical simulation (DNS), based on the Navier–Stokes equations, makes it possible to simulate turbulent flows at moderate Reynolds numbers. Restrictions depend on the power of the computer used and the efficiency of the solution algorithm. The results of DNS have been found to agree well with experimental data for some flows.

Most flows of interest have Reynolds numbers much too high for DNS to be a viable option, given the state of computational power for the next few decades. Any flight vehicle large enough to carry a human (L > 3 m), moving faster than 72 km/h (20 m/s) is well beyond the limit of DNS simulation (Re = 4 million). Transport aircraft wings (such as on an Airbus A300 or Boeing 747) have Reynolds numbers of 40 million (based on the wing chord). In order to solve these real-life flow problems, turbulence models will be a necessity for the foreseeable future. Reynolds-averaged Navier–Stokes equations (RANS) combined with turbulence modelling provides a model of the effects of the turbulent flow. Such a modelling mainly provides the additional momentum transfer by the Reynolds stresses, although the turbulence also enhances the heat and mass transfer. Another promising methodology is large eddy simulation (LES), especially in the guise of detached eddy simulation (DES)—which is a combination of RANS turbulence modelling and large eddy simulation.

Newtonian vs Non-Newtonian Fluids

Sir Isaac Newton showed how stress and the rate of strain are very close to linearly related for many familiar fluids, such as water and air. These Newtonian fluids are modelled by a coefficient called viscosity, which depends on the specific fluid.

However, some of the other materials, such as emulsions and slurries and some visco-elastic materials (e.g. blood, some polymers), have more complicated *non-Newtonian* stress-strain behaviours. These materials include *sticky liquids* such as latex, honey, and lubricants which are studied in the sub-discipline of rheology.

Subsonic vs Transonic, Supersonic and Hypersonic Flows

While many terrestrial flows (e.g. flow of water through a pipe) occur at low mach numbers, many flows of practical interest (e.g. in aerodynamics) occur at high fractions of the Mach Number M=1 or in excess of it (supersonic flows). New phenomena occur at these Mach number regimes (e.g. shock waves for supersonic flow, transonic instability in a regime of flows with M nearly equal to 1, non-equilibrium chemical behaviour due to ionization in hypersonic flows) and it is necessary to treat each of these flow regimes separately.

Magnetohydrodynamics

Magnetohydrodynamics (MHD) (*magneto fluid dynamics* or *hydromagnetics*) is the study of the dynamics of electrically conducting fluids. Examples of such fluids include plasmas, liquid metals, and salt water or electrolytes. The word *magnetohydrodynamics (MHD)* is derived from *magneto-* meaning magnetic field, *hydro-* meaning liquid, and *-dynamics* meaning movement. The field of MHD was initiated by Hannes Alfvén, for which he received the Nobel Prize in Physics in 1970. The fundamental concept behind MHD is that magnetic fields can induce currents in a moving conductive fluid, which in turn creates forces on the fluid and also changes the magnetic field itself. The set of equations which describe MHD are a combination of the Navier-Stokes equations of fluid dynamics and Maxwell's equations of electromagnetism. These differential equations have to be solved simultaneously, either analytically or numerically.

History

The first recorded use of the word *magnetohydrodynamics* is by Hannes Alfvén in 1942:

"At last some remarks are made about the transfer of momentum from the Sun to the planets, which is fundamental to the theory (§11). The importance of the magnetohydrodynamic waves in this respect are pointed out."

The ebbing salty water flowing past London's Waterloo Bridge interacts with the Earth's magnetic field to produce a potential difference between the two river-banks. Michael Faraday tried this experiment in 1832 but the current was too small to measure with the equipment at the time, and the river bed contributed to short-circuit the signal. However, by a similar process the voltage induced by the tide in the English Channel was measured in 1851.

Ideal and Resistive MHD

The simplest form of MHD, Ideal MHD, assumes that the fluid has so little resistivity that it can be treated as a perfect conductor. This is the limit of infinite magnetic Reynolds number. In ideal MHD, Lenz's law dictates that the fluid is in a sense *tied* to the magnetic field lines. To explain, in ideal MHD a small rope-like volume of fluid surrounding a field line will continue to lie along a magnetic field line, even as it is twisted and distorted by fluid flows in the system.

The connection between magnetic field lines and fluid in ideal MHD fixes the topology of the magnetic field in the fluid—for example, if a set of magnetic field lines are tied into a knot, then they will remain so as long as the fluid/plasma has negligible resistivity. This difficulty in reconnecting magnetic field lines makes it possible to store energy by moving the fluid or the source of the magnetic field. The energy can then become available if the conditions for ideal MHD break down, allowing magnetic reconnection that releases the stored energy from the magnetic field.

Ideal MHD Equations

The ideal MHD equations consist of the continuity equation, the Cauchy momentum equation, Ampere's Law neglecting displacement current, and a temperature evolution equation. As with any fluid description to a kinetic system, a closure approximation must be applied to highest moment of the particle distribution equation. This is often accomplished with approximations to the heat flux through a condition of adiabaticity or isothermality.

In the following, B is the magnetic field, E is the electric field, v is the bulk plasma velocity, J is the current density, ρ is the mass density, p is the plasma pressure, and t is time. The continuity equation is

$$\frac{\partial \rho}{\partial t} + \nabla \cdot (\rho \mathrm{v}) = 0.$$

The momentum equation is

$$\rho\left(\frac{\partial}{\partial t} + \mathrm{v}\cdot\nabla\right)\mathrm{v} = \mathrm{J}\times\mathrm{B} - \nabla p.$$

The Lorentz force term $\mathrm{J}\times\mathrm{B}$ can be expanded to give

$$\mathrm{J}\times\mathrm{B} = \frac{(\mathrm{B}\cdot\nabla)\mathrm{B}}{\mu_0} - \nabla\left(\frac{B^2}{2\mu_0}\right),$$

where the first term on the right hand side is the magnetic tension force and the second term is the magnetic pressure force. The ideal Ohm's law for a plasma is given by

$$\mathrm{E} + \mathrm{v}\times\mathrm{B} = 0.$$

Faraday's law is

$$\frac{\partial \mathrm{B}}{\partial t} = -\nabla\times\mathrm{E}.$$

The low-frequency Ampere's law neglects displacement current and is given by

$$\mu_0 \mathrm{J} = \nabla\times\mathrm{B}.$$

The magnetic divergence constraint is

$$\nabla\cdot\mathrm{B} = 0.$$

The energy equation is given by

$$\frac{\mathrm{d}}{\mathrm{d}t}\left(\frac{p}{\rho^{\gamma}}\right) - 0,$$

where $\gamma = 5/3$ is the ratio of specific heats for an adiabatic equation of state. This energy equation is, of course, only applicable in the absence of shocks or heat conduction as it assumes that the entropy of a fluid element does not change.

Applicability of Ideal MHD to Plasmas

Ideal MHD is only strictly applicable when:

1. The plasma is strongly collisional, so that the time scale of collisions is shorter than the other characteristic times in the system, and the particle distributions are therefore close to Maxwellian.
2. The resistivity due to these collisions is small. In particular, the typical magnetic diffusion times over any scale length present in the system must be longer than any time scale of interest.
3. Interest in length scales much longer than the ion skin depth and Larmor radius perpendicular to the field, long enough along the field to ignore Landau damping, and time scales much longer than the ion gyration time (system is smooth and slowly evolving).

Importance of Resistivity

In an imperfectly conducting fluid the magnetic field can generally move through the fluid following a diffusion law with the resistivity of the plasma serving as a diffusion constant. This means that solutions to the ideal MHD equations are only applicable for a limited time for a region of a given size before diffusion becomes too important to ignore. One can estimate the diffusion time across a solar active region (from collisional resistivity) to be hundreds to thousands of years, much longer than the actual lifetime of a sunspot—so it would seem reasonable to ignore the resistivity. By contrast, a metre-sized volume of seawater has a magnetic diffusion time measured in milliseconds.

Even in physical systems which are large and conductive enough that simple estimates of the Lundquist number suggest that we can ignore the resistivity, resistivity may still be important: many instabilities exist that can increase the effective resistivity of the plasma by factors of more than a billion. The enhanced resistivity is usually the result of the formation of small scale structure like current sheets or fine scale magnetic turbulence, introducing small spatial scales into the system over which ideal MHD is broken and magnetic diffusion can occur quickly. When this happens, magnetic reconnection may occur in the plasma to release stored magnetic energy as waves, bulk mechanical acceleration of material, particle acceleration, and heat. Magnetic reconnection in highly conductive systems is important because it concentrates energy in time and space, so that gentle forces applied to a plasma for long periods of time can cause violent explosions and bursts of radiation.

When the fluid cannot be considered as completely conductive, but the other conditions for ideal MHD are satisfied, it is possible to use an extended model called resistive MHD. This includes an extra term in Ohm's Law which models the collisional resistivity. Generally MHD computer simulations are at least somewhat resistive because their computational grid introduces a numerical resistivity.

Importance of Kinetic Effects

Another limitation of MHD (and fluid theories in general) is that they depend on the assumption that the plasma is strongly collisional (this is the first criterion listed above), so that the time scale of collisions is shorter than the other characteristic times in the system, and the particle distributions are Maxwellian. This is usually not the case in fusion, space and astrophysical plasmas. When this is not the case, or we are interested in smaller spatial scales, it may be necessary to use a kinetic model which properly accounts for the non-Maxwellian shape of the distribution function. However, because MHD is relatively simple and captures many of the important properties of plasma dynamics it is often qualitatively accurate and is almost invariably the first model tried. Effects which are essentially kinetic and not captured by fluid models include double layers, Landau damping, a wide range of instabilities, chemical separation in space plasmas and electron runaway.

Structures in MHD Systems

In many MHD systems most of the electric current is compressed into thin nearly-two-dimensional ribbons termed current sheets. These can divide the fluid into magnetic domains, inside of which the currents are relatively weak. Current sheets in the solar corona are thought to be between a few metres and a few kilometres in thickness, which is quite thin compared to the magnetic domains (which are thousands to hundreds of thousands of kilometres across). Another example is in the Earth's magnetosphere, where current sheets separate topologically distinct domains, isolating most of the Earth's ionosphere from the solar wind.

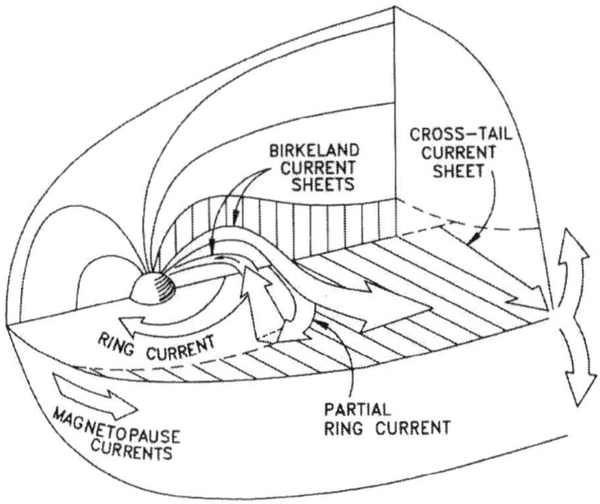

Figure: *Schematic view of the different current systems which shape the Earth's magnetosphere*

MHD Waves

The wave modes derived using MHD plasma theory are called magnetohydrodynamic waves or MHD waves. In general there are three MHD wave modes:

- Pure (or oblique) Alfvén wave
- Slow MHD wave
- Fast MHD wave

All these waves have constant phase velocities for all frequencies, and hence there is no dispersion. At the limits when the angle between the wave propagation vector k and magnetic field B is either 0 (180) or 90 degrees, the wave modes are called:

name	*type*	*propagation*	*phase velocity*	*association*	*medium*	*other names*
Sound wave	longitudinal	$\vec{k} \parallel \vec{B}$	adiabatic sound velocity	none	compressible, nonconducting fluid	
Alfvén wave	transverse	$\vec{k} \parallel \vec{B}$	Alfvén velocity	B		shear Alfvén wave, the slow Alfvén wave, torsional Alfvén wave
Magnetosonic wave	longitudinal	$\vec{k} \perp \vec{B}$		B, E		compressional Alfvén wave, fast Alfvén wave, magnetoacoustic wave

The MHD oscillations will be damped if the fluid is not perfectly conducting but has a finite conductivity, or if viscous effects are present.

MHD waves and oscillations are a popular tool for the remote diagnostics of laboratory and astrophysical plasmas, e.g. the corona of the Sun (Coronal seismology).

Extensions to Magnetohydrodynamics

Resistive MHD: Resistive MHD describes magnetized fluids with finite electron diffusivity ($\eta \neq 0$). This diffusivity leads to a breaking in the magnetic topology; magnetic field lines can 'reconnect' when they collide. Usually this term is small and reconnections can be handled by thinking of them as not dissimilar to shocks; this process has been shown to be important in the Earth-Solar magnetic interactions.

Extended MHD

Extended MHD describes a class of phenomena in plasmas that are higher order than resistive MHD, but which can adequately be treated with a single fluid description. These include the effects of Hall physics, electron pressure gradients, finite Larmor Radii in the particle gyromotion, and electron inertia.

Two-Fluid MHD

Two-Fluid MHD describes plasmas that include a non-negligible Hall electric field. As a result, the electron and ion momenta must be treated separately. This description is more closely tied to Maxwell's equations as an evolution equation for the electric field exists.

Hall MHD

In 1960, M. J. Lighthill criticized the applicability of ideal or resistive MHD theory for plasmas. It concerned the neglect of the "Hall current term", a frequent simplification made in magnetic fusion theory. Hall-magnetohydrodynamics (HMHD) takes into account this electric field description of magnetohydrodynamics. The most important difference is that in the absence of field line breaking, the magnetic field is tied to the electrons and not to the bulk fluid.

Collisionless MHD

MHD is also often used for collisionless plasmas. In that case the MHD equations are derived from the Vlasov equation.

Applications

Geophysics: Beneath the Earth's mantle, lies the core which is made up of two parts - the solid inner core and liquid outer core - both have significant quantities of iron. The liquid outer core moves in the presence of the magnetic field and eddies are set up into the same due to the Coriolis effect. These eddies develop a magnetic field which boosts Earth's original magnetic field - a process which is self-sustaining and is called as the geomagnetic dynamo.

Based on the MHD equations, Glatzmaier and Paul Roberts have made a supercomputer model of the Earth's interior. After running the simulations for thousands of years in virtual time, the changes in Earth's magnetic field can be studied. The simulation results are in good agreement with the observations as the simulations have correctly predicted that the Earth's magnetic field flips every few hundred thousands of years. During the flips, the magnetic field doesn't vanish altogether - it just gets more complicated.

Astrophysics

MHD applies quite well to astrophysics since over 99% of baryonic matter content of the Universe is made up of plasma, including stars, the interplanetary medium (space between the planets), the interstellar medium (space between the stars), the intergalactic medium, nebulae and jets. Many astrophysical systems are not in local thermal equilibrium, and therefore require an additional kinematic treatment to describe all the phenomena within the system.

Sunspots are caused by the Sun's magnetic fields, as Joseph Larmor theorized in 1919. The solar wind is also governed by MHD. The differential solar rotation may be the long term effect of magnetic drag at the poles of the Sun, an MHD phenomenon due to the Parker spiral shape assumed by the extended magnetic field of the Sun.

Previously, theories describing the formation of the Sun and planets could not explain how the Sun has 99.87% of the mass, yet only 0.54% of the angular momentum in the solar system. In a closed system such as the cloud of gas and dust from which the Sun was formed, mass and angular momentum are both conserved. That conservation would imply that as the mass concentrated in the centre of the cloud to form the Sun, it would spin up, much like a skater pulling their arms in. The high speed of rotation predicted by early theories would have flung the proto-Sun apart before it could have formed. However, magnetohydrodynamic effects transfer the Sun's angular momentum into the outer solar system, slowing its rotation.

Breakdown of ideal MHD (in the form of magnetic reconnection) is known to be the cause of solar flares, the largest explosions in the solar system. The magnetic field in a solar active region over a sunspot can become quite stressed over time, storing energy that is released suddenly as a burst of motion, X-rays, and radiation when the main current sheet collapses, reconnecting the field.

Engineering

MHD is related to engineering problems such as plasma confinement, liquid-metal cooling of nuclear reactors, and electromagnetic casting (among others). A magnetohydrodynamic drive or MHD propulsor is a method for propelling seagoing vessels using only electric and magnetic fields with no moving parts, using magnetohydrodynamics. The working principle involves electrification of the propellant (gas or water) which can then be directed by a magnetic field, pushing the vehicle in the opposite direction. Although some working prototypes exist, MHD drives remain impractical.

The first prototype of this kind of propulsion was built and tested in 1965 by Steward Way, a professor of mechanical engineering at the University of California, Santa Barbara. Way, on leave from his job at Westinghouse Electric, assigned his senior year undergraduate students to develop a submarine with this new propulsion system. In early 1990s, Mitsubishi built a boat, the 'Yamato,' which uses a magnetohydrodynamic drive, is driven by a liquid helium-cooled superconductor, and can travel at 15 km/h. MHD power generation fuelled by potassium-seeded coal combustion gas showed potential for more efficient energy conversion (the absence of solid moving parts allows operation at higher temperatures), but failed due to cost prohibitive technical difficulties. One major engineering problem was the failure of the primary coal combustion chamber's wall due to abrasion. In microfluidics, MHD is studied as a fluid pump for producing a continuous, nonpulsating flow in a complex microchannel design.

Magnetic Drug Targeting

An important task in cancer research is developing more precise methods for delivery of medicine to affected areas. One method involves the binding of medicine to biologically compatible magnetic particles (e.g. ferrofluids), which are guided to the target via careful placement of permanent magnets on the external body. Magnetohydrodynamic equations and finite element analysis are used to study the interaction between the magnetic fluid particles in the bloodstream and the external magnetic field.

Other Approximations

There are a large number of other possible approximations to fluid dynamic problems. Some of the more commonly used are listed below.

- The Boussinesq approximation neglects variations in density except to calculate buoyancy forces. It is often used in free convection problems where density changes are small.
- Lubrication theory and Hele-Shaw flow exploits the large aspect ratio of the domain to show that certain terms in the equations are small and so can be neglected.
- Slender-body theory is a methodology used in Stokes flow problems to estimate the force on, or flow field around, a long slender object in a viscous fluid.
- The shallow-water equations can be used to describe a layer of relatively inviscid fluid with a free surface, in which surface gradients are small.
- The Boussinesq equations are applicable to surface waves on thicker layers of fluid and with steeper surface slopes.
- Darcy's law is used for flow in porous media, and works with variables averaged over several pore-widths.
- In rotating systems, the quasi-geostrophic approximation assumes an almost perfect balance between pressure gradients and the Coriolis force. It is useful in the study of atmospheric dynamics.

Boussinesq Approximation (Buoyancy)

In fluid dynamics, the Boussinesq approximation (pronounced: [busin[Úsk], named for Joseph Valentin Boussinesq) is used in the field of buoyancy-driven flow (also known as natural convection). It states that density differences are sufficiently small to be neglected, except where they appear in terms multiplied by *g*, the acceleration due to gravity. The essence of the Boussinesq approximation is that the difference in inertia is negligible but gravity is sufficiently strong to make the specific weight appreciably different between the two fluids. Sound waves are impossible/neglected when the Boussinesq approximation is used since sound waves move via density variations.

Boussinesq flows are common in nature (such as atmospheric fronts, oceanic circulation, katabatic winds), industry (dense gas dispersion, fume cupboard ventilation), and the built environment (natural ventilation, central heating). The approximation is extremely accurate for many such flows, and makes the mathematics and physics simpler. The approximation's advantage arises because when considering a flow of, say, warm and cold water of density ρ_1 and ρ_2 one needs only consider a single density ρ: the difference $\Delta\rho = \rho_1 - \rho_2$ is negligible. Dimensional analysis shows that, under these circumstances, the only sensible way that acceleration due to gravity g should enter into the equations of motion is in the reduced gravity g' where

$$g' = g\frac{\rho_1 - \rho_2}{\rho}.$$

(Note that the denominator may be either density without affecting the result because the change would be of order $g(\Delta\rho / \rho)^2$). The most generally used dimensionless number would be the Richardson number and Rayleigh number. The mathematics of the flow is therefore simpler because the density ratio (ρ_1 / ρ_2, a dimensionless number) does not affect the flow; the Boussinesq approximation states that it may be assumed to be exactly one.

Inversions

One feature of Boussinesq flows is that they look the same when viewed upside-down, provided that the identities of the fluids are reversed. The Boussinesq approximation is *inaccurate* when the nondimensionalised density difference $\Delta\rho / \rho$ is of order unity. For example, consider an open window in a warm room. The warm air inside is lighter than the cold air outside, which flows into the room and down towards the floor. Now imagine the opposite: a cold room exposed to warm outside air. Here the air flowing in moves up toward the ceiling. If the flow is Boussinesq (and the room is otherwise symmetrical), then viewing the cold room upside down is exactly the same as viewing the warm room right-way-round. This is because the only way density enters the problem is via the reduced gravity g' which undergoes only a sign change when changing from the warm room flow to the cold room flow.

An example of a non-Boussinesq flow is bubbles rising in water. The behaviour of air bubbles rising in water is very different from the behaviour of water falling in air: in the former case rising bubbles tend to form hemispherical shells, while water falling in air splits into raindrops (at small length scales surface tension enters the problem and confuses the issue).

Chapter 2

Lubrication Theory

In fluid dynamics, lubrication theory describes the flow of fluids (liquids or gases) in a geometry in which one dimension is significantly smaller than the others.

Interior flows are those where the boundaries of the fluid volume are known, and include those inside bearings. Here a key goal of lubrication theory is to determine the pressure distribution in the fluid volume, and hence the forces on the bearing components. The working fluid in this case is often termed a lubricant.

Free film lubrication theory is concerned with the case in which one of the surfaces containing the fluid is a free surface. In that case the position of the free surface is itself unknown, and one goal of lubrication theory is then to determine this. Surface tension may then be significant, or even dominant. Issues of wetting and dewetting then arise. For very thin films (thickness less than one micrometre), additional intermolecular forces, such as disjoining forces, may become significant.

Theoretical Basis

Mathematically, lubrication theory can be seen as exploiting the disparity between two length scales. The first is the characteristic film thickness, *H*, and the second is a characteristic substrate length scale *L*. The key requirement for lubrication theory is that the ratio $\epsilon = H / L$ is small, that is, $\epsilon \ll 1$. The Navier-Stokes equations (or Stokes equations, when fluid inertia may be neglected) are expanded in this small parameter, and the leading-order equations are then

$$\frac{\partial p}{\partial z} = 0$$

$$\frac{\partial p}{\partial x} = \mu \frac{\partial^2 u}{\partial z^2}$$

where x and z are coordinates in the direction of the substrate and perpendicular to it respectively. Here p is the fluid pressure, and u is the fluid velocity component parallel to the substrate; μ is the fluid viscosity.

Further details can be found in the literature or in the textbooks given in the bibliography.

Applications

An important application area is lubrication of machinery components such as fluid bearings and mechanical seals. Coating is another major application area including the preparation of thin films, printing, painting and adhesives.

Biological applications have included studies of red blood cells in narrow capillaries and of liquid flow in the lung and eye.

Hele-Shaw Flow

Hele-Shaw flow (named after Henry Selby Hele-Shaw) is defined as Stokes flow between two parallel flat plates separated by an infinitesimally small gap. Various problems in fluid mechanics can be approximated to Hele-Shaw flows and thus the research of these flows is of importance. Approximation to Hele-Shaw flow is specifically important to micro-flows. This is due to manufacturing techniques, which creates shallow planar configurations, and the typically low Reynolds numbers of micro-flows.

The governing equation of Hele-Shaw flows is identical to that of the inviscid potential flow and to the flow of fluid through a porous medium (Darcy's law). It thus permits visualization of this kind of flow in two dimensions.

Mathematical Formulation of Hele-Shaw Flows

Let x,y be the directions parallel to the flat plates, and the perpendicular direction, with $2H$ being the gap between the plates (at $z = \pm H$). When the gap between plates is asymptotically small

$$H \to 0,$$

the velocity profile in the z direction is parabolic (i.e. is a quadratic function of the coordinate in this direction). The equation relating the pressure gradient to the velocity is,

$$u = \nabla p \frac{z^2 - H^2}{2\mu}$$

where u is the velocity $p(x,y,t)$, is the local pressure, μ is the fluid viscosity.

This relation and the uniformity of the pressure in the narrow direction permits us to integrate the velocity with regard to and thus to consider an effective velocity field in only the two dimensions x and y. When substituting this equation into the continuity equation and integrating over z we obtain the governing equation of Hele-Shaw flows,

$$\frac{\partial^2 p}{\partial x^2} + \frac{\partial^2 p}{\partial y^2} = 0.$$

This equation is supplemented by the no-penetration boundary conditions on the side walls of the geometry,

$$\nabla p \cdot \hat{n} = 0$$

where $\hat{n}$ is a unit vector perpendicular to the side wall.

Hele-Shaw Cell

The term Hele-Shaw cell is commonly used for cases in which a fluid is injected into the shallow geometry from above or below the geometry, and when the fluid is bounded by another liquid or gas. For such flows the boundary conditions are defined by pressures and surface tensions.

Slender-Body Theory

In fluid dynamics and electrostatics, slender-body theory is a methodology that can be used to take advantage of the slenderness of a body to obtain an approximation to a field surrounding it and/or the net effect of the field on the body. Principal applications are to Stokes flow — at very low Reynolds numbers — and in electrostatics.

Theory for Stokes Flow

Consider slender body of length ℓ and typical diameter $2a$ with $\ell \gg a$, surrounded by fluid of viscosity μ whose motion is governed by the Stokes equations. Note that Stokes paradox implies that the limit of infinite aspect ratio $\ell / a \to \infty$ is singular, as no Stokes flow can exist around an infinite cylinder.

Slender-body theory allows us to derive an approximate relationship between the velocity of the body at each point along its length and the force per unit length experienced by the body at that point.

Let the axis of the body be described by $X(s,t)$, where s is an arc-length coordinate, and t is time. By virtue of the slenderness of the body, the force exerted on the fluid at the surface of the body may be approximated by a distribution of Stokeslets along the axis with force density $f(s)$per unit length. $f$is assumed to vary only over lengths much greater than a , and the fluid velocity at the surface adjacent to $X(s,t)$ is well-approximated by $\partial X / \partial t$.

The fluid velocity u(x) at a general point x due to such a distribution can be written in terms of an integral of the Oseen tensor (named after Carl Wilhelm Oseen), which acts as a Greens function for a single Stokeslet. We have

$$u(x) = \int_0^{\ell} \frac{f(s)}{8\pi\mu} \cdot \left(\frac{I}{|x - X|} + \frac{(x - X)(x - X)}{|x - X|^3} \right) ds$$

where I is the identity tensor.

Asymptotic analysis can then be used to show that the leading-order contribution to the integral for a point x on the surface of the body adjacent to position s_0 comes from the force distribution at $|s - s_0| = O(a)$. Since $a \ll \ell$, we approximate $f(s) \approx f(s_0)$. We then obtain

$$\frac{\partial X}{\partial t} \sim \frac{\ln(\ell / a)}{4\pi\mu} f(s) \cdot \left(I + X'X' \right)$$

where $X' = \partial X / \partial s$.

The expression may be inverted to give the force density in terms of the motion of the body:

$$f(s) \sim \frac{4\pi\mu}{\ln(\ell / a)} \frac{\partial X}{\partial t} \cdot \left(I - \frac{1}{2} X'X' \right)$$

Two canonical results that follow immediately are for the drag force F on a rigid cylinder (length ℓ, radius a) moving a velocity u either parallel to its axis or perpendicular to it. The parallel case gives

$$F \sim \frac{2\pi\mu\ell u}{\ln(\ell / a)}$$

while the perpendicular case gives

$$F \sim \frac{4\pi\mu\ell u}{\ln(\ell / a)}$$

with only a factor of two difference.

Note that the dominant length scale in the above expressions is the longer length ℓ; the shorter length has only a weak effect through the logarithm of the aspect ratio. In slender-body theory results, there are $O(1)$ corrections to the logarithm, so even for relatively large values of ℓ / a the error terms will not be that small.

Boussinesq Approximation (Water Waves)

In fluid dynamics, the Boussinesq approximation for water waves is an approximation valid for weakly non-linear and fairly long waves. The approximation is named after Joseph Boussinesq, who first derived them in response to the observation by John Scott Russell of the wave of translation (also known as solitary wave or soliton). The 1872 paper of Boussinesq introduces the equations now known as the Boussinesq equations.

The Boussinesq approximation for water waves takes into account the vertical structure of the horizontal and vertical flow velocity. This results in non-linear partial differential equations, called Boussinesq-type equations, which incorporate frequency dispersion (as opposite to the shallow water equations, which are not frequency-dispersive). In coastal engineering, Boussinesq-type equations are frequently used in computer models for the simulation of water waves in shallow seas and harbours.

While the Boussinesq approximation is applicable to fairly long waves – that is, when the wavelength is large compared to the water depth – the Stokes expansion is more appropriate for short waves (when the wavelength is of the same order as the water depth, or shorter).

Boussinesq Approximation

The essential idea in the Boussinesq approximation is the elimination of the vertical coordinate from the flow equations, while retaining some of the influences of the vertical structure of the flow under water waves. This is useful because the waves propagate in the horizontal plane and have a different (not wave-like) behaviour in the vertical direction. Often, as in Boussinesq's case, the interest is primarily in the wave propagation.

This elimination of the vertical coordinate was first done by Joseph Boussinesq in 1871, to construct an approximate solution for the solitary wave (or wave of translation). Subsequently, in 1872, Boussinesq derived the equations known nowadays as the Boussinesq equations.

The steps in the Boussinesq approximation are:

- a Taylor expansion is made of the horizontal and vertical flow velocity (or velocity potential) around a certain elevation,
- this Taylor expansion is truncated to a finite number of terms,
- the conservation of mass for an incompressible flow and the zero-curl condition for an irrotational flow are used, to replace vertical partial derivatives of quantities in the Taylor expansion with horizontal partial derivatives.

Thereafter, the Boussinesq approximation is applied to the remaining flow equations, in order to eliminate the dependence on the vertical coordinate. As a result, the resulting partial differential equations are in terms of functions of the horizontal coordinates (and time). As an example, consider potential flow over a horizontal bed in the (x,z) plane, with x the horizontal and z the vertical coordinate. The bed is located at $z=-h$, where h is the mean water depth. A Taylor expansion is made of the velocity potential $\varphi(x,z,t)$ around the bed level $z = -h$:

$$\begin{aligned}\varphi = \; & \varphi_b + (z+h)\left[\frac{\partial\varphi}{\partial z}\right]_{z=-h} + \frac{1}{2}(z+h)^2\left[\frac{\partial^2\varphi}{\partial z^2}\right]_{z=-h} \\ & + \frac{1}{6}(z+h)^3\left[\frac{\partial^3\varphi}{\partial z^3}\right]_{z=-h} + \frac{1}{24}(z+h)^4\left[\frac{\partial^4\varphi}{\partial z^4}\right]_{z=-h} + \cdots,\end{aligned}$$

where $\varphi_b(x,t)$ is the velocity potential at the bed. Invoking Laplace's equation for φ, as valid for incompressible flow, gives:

$$\begin{aligned}\varphi = \; & \left\{\varphi_b - \frac{1}{2}(z+h)^2\frac{\partial^2\varphi_b}{\partial x^2} + \frac{1}{24}(z+h)^4\frac{\partial^4\varphi_b}{\partial x^4} + \cdots\right\} \\ & + \left\{(z+h)\left[\frac{\partial\varphi}{\partial z}\right]_{z=-h} - \frac{1}{6}(z+h)^3\frac{\partial^2}{\partial x^2}\left[\frac{\partial\varphi}{\partial z}\right]_{z=-h} + \cdots\right\} \\ = \; & \left\{\varphi_b - \frac{1}{2}(z+h)^2\frac{\partial^2\varphi_b}{\partial x^2} + \frac{1}{24}(z+h)^4\frac{\partial^4\varphi_b}{\partial x^4} + \cdots\right\},\end{aligned}$$

since the vertical velocity $\partial\varphi / \partial z$ is zero at the – impermeable – horizontal bed $z = -h$. This series may subsequently be truncated to a finite number of terms.

Original Boussinesq Equations

Derivation: For water waves on an incompressible fluid and irrotational flow in the (x,z) plane, the boundary conditions at the free surface elevation $z = \eta(x,t)$ are:

$$\frac{\partial \eta}{\partial t}+u\frac{\partial \eta}{\partial x}-w=0$$

$$\frac{\partial \varphi}{\partial t}+\frac{1}{2}\left(u^2+w^2\right)+g\eta=0,$$

where:

u is the horizontal flow velocity component: $u = \partial\varphi / \partial x$,

w is the vertical flow velocity component: $w = \partial\varphi / \partial z$,

g is the acceleration by gravity.

Now the Boussinesq approximation for the velocity potential φ, as given above, is applied in these boundary conditions. Further, in the resulting equations only the linear and quadratic terms with respect to η and u_b are retained (with $u_b = \partial\varphi_b / \partial x$ the horizontal velocity at the bed $z = -h$). The cubic and higher order terms are assumed to be negligible. Then, the following partial differential equations are obtained:

set A – Boussinesq (1872), equation (25)

$$\frac{\partial \eta}{\partial t}+\frac{\partial}{\partial x}\left[(h+\eta)u_b\right]=\frac{1}{6}h^3\frac{\partial^3 u_b}{\partial x^3},$$

$$\frac{\partial u_b}{\partial t}+u_b\frac{\partial u_b}{\partial x}+g\frac{\partial \eta}{\partial x}=\frac{1}{2}h^2\frac{\partial^3 u_b}{\partial t\partial x^2}.$$

This set of equations has been derived for a flat horizontal bed, *i.e.* the mean depth h is a constant independent of position x. When the right-hand sides of the above equations are set to zero, they reduce to the shallow water equations.

Under some additional approximations, but at the same order of accuracy, the above set A can be reduced to a single partial differential equation for the free surface elevation η:

set B – Boussinesq (1872), equation (26)

$$\frac{\partial^2 \eta}{\partial t^2}-gh\frac{\partial^2 \eta}{\partial x^2}-gh\frac{\partial^2}{\partial x^2}\left(\frac{3}{2}\frac{\eta^2}{h}+\frac{1}{3}h^2\frac{\partial^2 \eta}{\partial x^2}\right)=0.$$

From the terms between brackets, the importance of nonlinearity of the equation can be expressed in terms of the Ursell number. In dimensionless quantities, using the water depth h and gravitational acceleration g for non-dimensionalization, this equation reads, after normalization:

$$\frac{\partial^2\psi}{\partial\tau^2}-\frac{\partial^2\psi}{\partial\xi^2}-\frac{\partial^2}{\partial\xi^2}\left(\frac{1}{2}\psi^2+\frac{\partial^2\psi}{\partial\xi^2}\right)=0,$$

with:

$\psi=3\frac{\eta}{h}$	: the dimensionless surface elevation,
$\tau=\sqrt{3}t\sqrt{\frac{g}{h}}$	: the dimensionless time, and
$\xi=\sqrt{3}\frac{x}{h}$	: the dimensionless horizontal position.

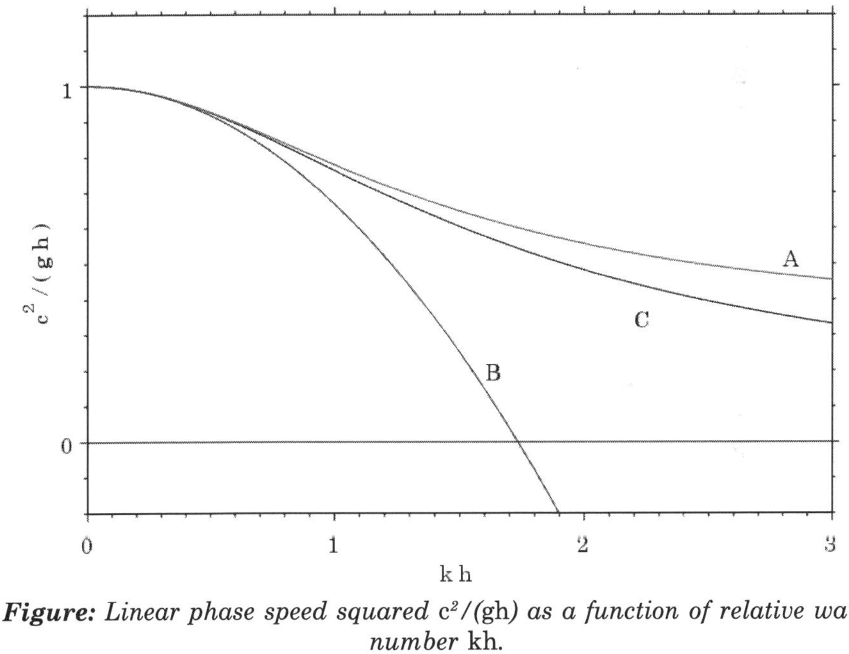

***Figure:** Linear phase speed squared* c²/(gh) *as a function of relative wave number* kh.

A = Boussinesq (1872), equation (25),

B = Boussinesq (1872), equation (26),

C = full linear wave theory

Linear Frequency Dispersion

Water waves of different wave lengths travel with different phase speeds, a phenomenon known as frequency dispersion. For the case of infinitesimal wave amplitude, the terminology is *linear frequency dispersion.* The frequency dispersion characteristics of a Boussinesq-type of equation can be used to determine the range of wave lengths, for which it is a valid approximation.

The linear frequency dispersion characteristics for the above set A of equations are:

$$c^2 = gh\frac{1+\frac{1}{6}k^2h^2}{1+\frac{1}{2}k^2h^2},$$

with:

- c the phase speed,
- k the wave number ($k = 2\pi / \lambda$, with λ the wave length).

The relative error in the phase speed c for set A, as compared with linear theory for water waves, is less than 4% for a relative wave number $kh < ½\ \pi$. So, in engineering applications, set A is valid for wavelengths λ larger than 4 times the water depth h.

The linear frequency dispersion characteristics of equation B are:

$$c^2 = gh\left(1-\frac{1}{3}k^2h^2\right).$$

The relative error in the phase speed for equation B is less than 4% for $kh < 2\pi/7$, equivalent to wave lengths λ longer than 7 times the water depth h, called fairly long waves.

For short waves with $k^2\ h^2 > 3$ equation B become physically meaningless, because there are no longer real-valued solutions of the phase speed. The original set of two partial differential equations (Boussinesq, 1872, equation 25) does not have this shortcoming. The shallow water equations have a relative error in the phase speed less than 4% for wave lengths λ in excess of 13 times the water depth h.

Extensions

There are an overwhelming number of mathematical models which are referred to as Boussinesq equations. This may easily lead to confusion, since often they are loosely referenced to as *the* Boussinesq equations, while in fact a variant thereof is considered. So it is more appropriate to call them Boussinesq-type equations. Strictly speaking, *the* Boussinesq equations is the above mentioned set B, since it is used in the analysis in the remainder of his 1872 paper. Some directions, into which the Boussinesq equations have been extended, are:

- varying bathymetry,

- improved frequency dispersion,
- improved non-linear behaviour,
- making a Taylor expansion around different vertical elevations,
- dividing the fluid domain in layers, and applying the Boussinesq approximation in each layer separately,
- inclusion of wave breaking,
- inclusion of surface tension,
- extension to internal waves on an interface between fluid domains of different mass density,
- derivation from a variational principle.

Further Approximations for One-Way Wave Propagation

While the Boussinesq equations allow for waves travelling simultaneously in opposing directions, it is often advantageous to only consider waves travelling in one direction. Under small additional assumptions, the Boussinesq equations reduce to:

- the Korteweg–de Vries equation for wave propagation in one horizontal dimension,
- the Kadomtsev–Petviashvili equation for wave propagation in two horizontal dimensions,
- the nonlinear Schrödinger equation (NLS equation) for the complex valued amplitude of narrowband waves (slowly modulated waves).

Besides solitary wave solutions, the Korteweg–de Vries equation also has periodic and exact solutions, called cnoidal waves. These are approximate solutions of the Boussinesq equation.

Shallow Water Equations

The shallow water equations (also called Saint Venant equations in its unidimensional form, after Adhémar Jean Claude Barré de Saint-Venant) are a set of hyperbolic partial differential equations (or parabolic if viscous shear is considered) that describe the flow below a pressure surface in a fluid (sometimes, but not necessarily, a free surface). The shallow water equations can also be simplified to the commonly used 1-D Saint Venant Equation.

The equations are derived from depth-integrating the Navier–Stokes equations, in the case where the horizontal length scale is much greater than the vertical length scale. Under this condition, conservation of mass implies that the vertical velocity of the fluid is small. It can be shown from the momentum equation that vertical pressure gradients are nearly hydrostatic, and that horizontal pressure gradients are due to the displacement of the pressure surface, implying that the horizontal velocity field is constant throughout the depth of the fluid. Vertically integrating allows the vertical velocity to be removed from the equations. The shallow water equations are thus derived.

While a vertical velocity term is not present in the shallow water equations, note that this velocity is not necessarily zero. This is an important distinction because, for example, the vertical velocity cannot be zero when the floor changes depth, and thus if it were zero only flat floors would be usable with the shallow water equations. Once a solution (i.e. the horizontal velocities and free surface displacement) has been found, the vertical velocity can be recovered via the continuity equation.

Situations in fluid dynamics where the horizontal length scale is much greater than the vertical length scale are common, so the shallow water equations are widely applicable. They are used with Coriolis forces in atmospheric and oceanic modelling, as a simplification of the primitive equations of atmospheric flow.

Shallow water equation models have only one vertical level, so they cannot directly encompass any factor that varies with height. However, in cases where the mean state is sufficiently simple, the vertical variations can be separated from the horizontal and several sets of shallow water equations can describe the state.

Equations

Conservative Form: The shallow water equations are derived from equations of conservation of mass and conservation of momentum (the Navier–Stokes equations), which hold even when the assumptions of shallow water break down, such as across a hydraulic jump. In the case of no Coriolis, frictional or viscous forces, the shallow-water equations are:

$$\frac{\partial \eta}{\partial t} + \frac{\partial(\eta u)}{\partial x} + \frac{\partial(\eta v)}{\partial y} = 0$$

$$\frac{\partial(\eta u)}{\partial t} + \frac{\partial}{\partial x}\left(\eta u^2 + \frac{1}{2} g \eta^2\right) + \frac{\partial(\eta u v)}{\partial y} = 0$$

$$\frac{\partial(\eta v)}{\partial t} + \frac{\partial(\eta u v)}{\partial x} + \frac{\partial}{\partial y}\left(\eta v^2 + \frac{1}{2} g \eta^2\right) = 0.$$

Here η is the total fluid column height. The 2D vector *(u,v)* is the fluid's horizontal velocity, averaged across the vertical column. *g* is acceleration due to gravity. The first equation is derived from mass conservation, the second two from momentum conservation.

Non-Conservative Form

The equations can also be written in terms of velocities instead of momenta. Since velocities are not subject to a fundamental conservation equation, in this form the equations do not hold across a shock or hydraulic jump:

$$\begin{aligned}
\frac{\partial u}{\partial t}+u\frac{\partial u}{\partial x}+v\frac{\partial u}{\partial y}-fv &= -g\frac{\partial h}{\partial x}-bu,\\
\frac{\partial v}{\partial t}+u\frac{\partial v}{\partial x}+v\frac{\partial v}{\partial y}+fu &= -g\frac{\partial h}{\partial y}-bv,\\
\frac{\partial h}{\partial t} &= -\frac{\partial}{\partial x}\big(u(H+h)\big)-\frac{\partial}{\partial y}\big(v(H+h)\big),
\end{aligned}$$

Ignoring bulk advection (u and v are small), and assuming the wave height is a small proportion of the mean height (h ≪ H), we have:

$$\begin{aligned}
\frac{\partial u}{\partial t}-fv &= -g\frac{\partial h}{\partial x}-bu,\\
\frac{\partial v}{\partial t}+fu &= -g\frac{\partial h}{\partial y}-bv,\\
\frac{\partial h}{\partial t} &= -H\Big(\frac{\partial u}{\partial x}+\frac{\partial v}{\partial y}\Big)
\end{aligned}$$

These equations use the following symbols:

u	is the velocity in the x direction, or zonal velocity
v	is the velocity in the y direction, or meridional velocity
h	is the height deviation of the horizontal pressure surface from its mean height H
H	is the mean height of the horizontal pressure surface
g	is the acceleration due to gravity
f	is the Coriolis coefficient associated with the Coriolis force, on Earth equal to 2Ω sin(φ), where Ω is the angular rotation rate of the Earth (π/12 radians/hour), and φ is the latitude
b	is the viscous drag coefficient

Wave Modelling by Shallow Water Equations

Shallow water equations can be used to model Rossby and Kelvin waves in the atmosphere, rivers, lakes and oceans as well as gravity waves in a smaller domain (e.g. surface waves in a bath). In order for shallow water equations to be valid, the wave length of the phenomenon they are supposed to model has to be much higher than the depth of the basin where the phenomenon takes place. Shallow water equations are especially suitable to model tides which have very large length scales (over hundred of kilometres). For tidal motion, even a very deep ocean may be considered as shallow as its depth will always be much smaller than the tidal wave length.

Darcy's Law

Darcy's law is a phenomenologically derived constitutive equation that describes the flow of a fluid through a porous medium. The law was formulated by Henry Darcy based on the results of experiments on the flow of water through beds of sand. It also forms the scientific basis of fluid permeability used in the earth sciences, particularly in hydrogeology.

Background

Although Darcy's law (an expression of conservation of momentum) was determined experimentally by Darcy, it has since been derived from the Navier-Stokes equations via homogenization. It is analogous to Fourier's law in the field of heat conduction, Ohm's law in the field of electrical networks, or Fick's law in diffusion theory. One application of Darcy's law is to water flow through an aquifer; Darcy's law along with the equation of conservation of mass are equivalent to the groundwater flow equation, one of the basic relationships of hydrogeology. Darcy's law is also used to describe oil, water, and gas flows through petroleum reservoirs.

Description

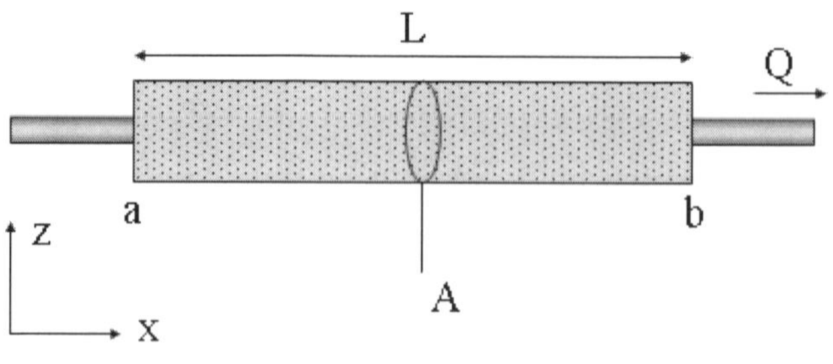

Figure: *Diagram showing definitions and directions for Darcy's law.*

Darcy's law is a simple proportional relationship between the instantaneous discharge rate through a porous medium, the viscosity of the fluid and the pressure drop over a given distance.

$$Q = \frac{-kA}{\mu} \frac{(P_b - P_a)}{L},$$

The total discharge, Q (units of volume per time, e.g., m^3/s) is equal to the product of the intrinsic permeability of the medium, k (m^2), the cross-sectional area to flow, A (units of area, e.g., m^2), and the pressure drop (Pb - Pa), all divided by the viscosity, μ (Pa ·s) and the length over which the pressure drop is taking place (m). The negative sign is needed because fluid flows from high pressure to low pressure. If the change in pressure is negative (where $P_a > P_b$), then the flow will be in the positive 'x' direction. Dividing both sides of the equation by the area and using more general notation leads to

$$q = \frac{-k}{\mu} \nabla P$$

where q is the flux (discharge per unit area, with units of length per time, m/s) and ∇P is the pressure gradient vector (Pa/m). This value of flux, often referred to as the Darcy flux, is not the velocity which the fluid travelling through the pores is experiencing. The fluid velocity (v) is related to the Darcy flux (q) by the porosity (n). The flux is divided by porosity to account for the fact that only a fraction of the total formation volume is available for flow. The fluid velocity would be the velocity a conservative tracer would experience if carried by the fluid through the formation.

$$v = \frac{q}{n}$$

Darcy's law is a simple mathematical statement which neatly summarizes several familiar properties that groundwater flowing in aquifers exhibits, including:

- if there is no pressure gradient over a distance, no flow occurs (these are hydrostatic conditions),
- if there is a pressure gradient, flow will occur from high pressure towards low pressure (opposite the direction of increasing gradient - hence the negative sign in Darcy's law),
- the greater the pressure gradient (through the same formation material), the greater the discharge rate, and
- the discharge rate of fluid will often be different through different formation materials (or even through the same

material, in a different direction) — even if the same pressure gradient exists in both cases.

A graphical illustration of the use of the steady-state groundwater flow equation (based on Darcy's law and the conservation of mass) is in the construction of flownets, to quantify the amount of groundwater flowing under a dam.

Darcy's law is only valid for slow, viscous flow; fortunately, most groundwater flow cases fall in this category. Typically any flow with a Reynolds number less than one is clearly laminar, and it would be valid to apply Darcy's law. Experimental tests have shown that flow regimes with Reynolds numbers up to 10 may still be Darcian, as in the case of groundwater flow. The Reynolds number (a dimensionless parameter) for porous media flow is typically expressed as

$$Re = \frac{\rho v d_{30}}{\mu}.$$

where ρ is the density of water (units of mass per volume), v is the specific discharge (not the pore velocity — with units of length per time), d_{30} is a representative grain diameter for the porous media (often taken as the 30% passing size from a grain size analysis using sieves - with units of length), and μ is the viscosity of the fluid.

Derivation

For stationary, creeping, incompressible flow, i.e. $D(\rho u_i)/Dt \approx 0$, the Navier-Stokes equation simplify to the Stokes equation:

$$\mu \nabla^2 u_i + \rho g_i - \partial_i P = 0,$$

where μ is the viscosity, u_i is the velocity in the i direction, g_i is the gravity component in the i direction and P is the pressure. Assuming the viscous resisting force is linear with the velocity we may write:

$$-\left(k_{ij}\right)^{-1} \mu \phi u_j + \rho g_i - \partial_i P = 0,$$

where ϕ is the porosity, and k_{ij} is the second order permeability tensor. This gives the velocity in the *n* direction,

$$k_{ni}\left(k_{ij}\right)^{-1} u_j = \delta_{nj} u_j = u_n = -\frac{k_{ni}}{\phi\mu}\left(\partial_i P - \rho g_i\right),$$

which gives Darcy's law for the volumetrix flux density in the direction,.

$$q_n = -\frac{k_{ni}}{\mu}\left(\partial_i P - \rho g_i\right)$$

In isotropic porous media the off-diagonal elements in the permeability tensor are zero $k_{ij} = 0$, for $i \neq j$ and the diagonal elements are identical, $k = k_{ii}$, and the common form is obtained

$$q = -\frac{k}{\mu}(\nabla P - \rho g).$$

Additional Forms of Darcy's Law

For very short time scales, a time derivative of flux may be added to Darcy's law, which results in valid solutions at very small times (in heat transfer, this is called the modified form of Fourier's law),

$$\tau \frac{\partial q}{\partial t} + q = -K\nabla h,$$

where τ is a very small time constant which causes this equation to reduce to the normal form of Darcy's law at "normal" times (> nanoseconds). The main reason for doing this is that the regular groundwater flow equation (diffusion equation) leads to singularities at constant head boundaries at very small times. This form is more mathematically rigorous, but leads to a hyperbolic groundwater flow equation, which is more difficult to solve and is only useful at very small times, typically out of the realm of practical use.

Another extension to the traditional form of Darcy's law is the Brinkman term, which is used to account for transitional flow between boundaries (introduced by Brinkman in 1949),

$$\beta \nabla^2 q + q = -K\nabla P,$$

where β is an effective viscosity term. This correction term accounts for flow through medium where the grains of the media are porous themselves, but is difficult to use, and is typically neglected.

Another derivation of Darcy's law is used extensively in petroleum engineering to determine the flow through permeable media - the most simple of which is for a one dimensional, homogeneous rock formation with a fluid of constant viscosity.

$$Q = \frac{kA}{\mu}\left(\frac{\partial P}{\partial L}\right),$$

where Q is the flowrate of the formation (in units of volume per unit time), k is the relative permeability of the formation (typically in millidarcies), A is the cross-sectional area of the formation, μ is the viscosity of the fluid (typically in units of centipoise, and L is the length of the porous media the fluid will flow through. $\partial P / \partial L$ represents the pressure change per unit length of the formation. This equation can also be solved for permeability, allowing for relative permeability to be calculated by forcing a fluid of known viscosity through a core of a known length and area, and measuring the pressure drop across the length of the core.

For very high velocities in porous media, inertial effects can also become significant. Sometimes an inertial term is added to the Darcy's equation, known as Forchheimer term. This term is able to account for the non-linear behaviour of the pressure difference vs velocity data.

$$\nabla P = -\frac{\mu}{k}q - \frac{\rho}{k_1}q^2,$$

where the additional term k_1 is known as inertial permeability.

Darcy's law is valid only for flow in continuum region. For a flow in transition region, where both viscous and Knudsen friction are present a new formulation is used, which is known as binary friction model

$$\nabla P = -\left(\frac{k}{\mu} + D_K\right)^{-1} q,$$

where D_K is the Knudsen diffusivity of the fluid in porous media.

Balanced Flow

In atmospheric science, balanced flow is an idealisation of atmospheric motion. The idealisation consists in considering the behaviour of one isolated parcel of air having constant density, its motion on a horizontal plane subject to selected forces acting on it and, finally, steady-state conditions.

Balanced flow is often an accurate approximation of the actual flow, and is useful in improving the qualitative understanding and interpretation of atmospheric motion. In particular, the balanced-flow speeds can be used as estimates of the wind speed for particular arrangements of the atmospheric pressure on Earth's surface.

The Momentum Equations in Natural Coordinates

Trajectories: The momentum equations are written for the generic trajectory of a packet of flow travelling on a horizontal plane and taken at a certain elapsed time *t*. The position of the packet is defined by the distance on the trajectory s=$s(t)$ that it has travelled by time

t. In reality, the trajectory is the outcome of the balance of forces upon the particle. In this section we assume to know it from the start for convenience of representation. When we consider the motion determined by the forces selected next, we will have clues of which type of trajectory fits the particular balance of forces.

The trajectory at a position s has one tangent unit vector s that invariably points in the direction of growing s's, as well as one unit vector n, perpendicular to s, that points towards the local centre of curvature O. The centre of curvature is found on the 'inner side' of the bend, and can shift across either side of the trajectory according to the shape of it. The distance between the parcel position and the centre of curvature is the radius of curvature R at that position. The radius of curvature approaches an infinite length at the points where the trajectory becomes straight and the positive orientation of n is not determined in this particular case (discussed in geostrophic flows). The frame of reference (s,n) is shown by the red arrows in the figure. This frame is termed natural or intrinsic because the axes continuously adjust to the moving parcel, and so they are the most closely connected to its fate.

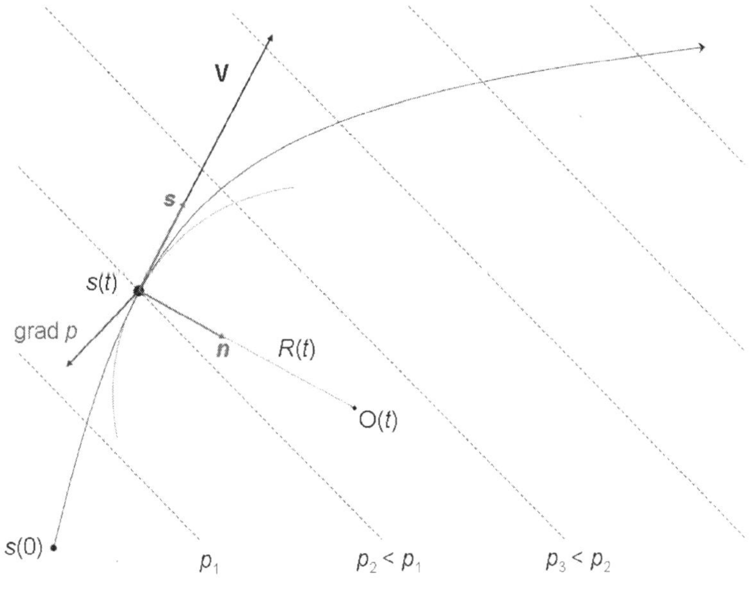

Kinematics

The velocity vector (V) is oriented like s and has intensity (speed) $V = ds/dt$. This speed is always a positive quantity, since any parcel moves along its own trajectory and, for increasing times ($dt>0$), the trodden length increases as well ($ds>0$).

The acceleration vector of the parcel is decomposed in the *tangential* acceleration parallel to s and in the *centripetal* acceleration along positive n. The tangential acceleration only changes the speed V and is equal to DV/Dt, where big d's denote the material derivative. The centripetal acceleration always points towards the centre of curvature O and only changes the direction s of the forward displacement while the parcel moves on.

Forces

In the balanced-flow idealization we consider a three-way balance of vector forces that are:

- Pressure force. This is the action on the parcel arising from the spatial differences of atmospheric pressure p around it. (Temporal changes are of no interest here.) The spatial change of pressure is visualised through isobars, that are contours joining the locations where the pressure has a same value. In the figure this is simplistically shown by equally spaced straight lines. The *pressure force* acting on the parcel is minus the gradient vector of p (in symbols: grad p) - drawn in the figure as a blue arrow. At all points, the pressure gradient points to the direction of maximum increase of p and is always normal to the isobar at that point. Since the flow packet feels a push from the higher to the lower pressures, the effective pressure vector force is contrary to the pressure gradient, whence the minus sign before the gradient vector.
- Friction. This is a force always opposing the forward motion, whereby the vector invariably acts in the negative direction s with an effect to reduce the speed. The friction at play in the balanced-flow models is the one exerted by the roughness of the Earth's surface on the air moving higher above. For simplicity, we here assume that the frictional force (per unit mass) adjusts to the parcel's speed proportionally through a constant coefficient of friction K. In more realistic conditions, the dependence of friction on the speed is non-linear except for slow laminar flows.
- Coriolis force. This action, due to the Earth's rotation, tends to displace any body travelling in the northern (southern) hemisphere towards its right (left). Its intensity per unit mass is proportional to the speed V and increases in magnitude from

the equator (where it is zero) towards the poles proportionally to the local Coriolis frequency f (a positive number north of the equator and negative south). Therefore, the Coriolis vector invariably points sideways, that is along the n axis. Its sign in the balance equation may change, since the positive orientation of n flips between right and left of the trajectory based solely on its curvature, while the Coriolis vector points to either side based on the packet's position on the Earth. The exact expression of the Coriolis force is a bit more complex than the product of the Coriolis parameter and parcel's velocity. However, this approximation is consistent with having neglected the curvature of the Earth's surface.

In the fictitious situation drawn in the figure, the pressure force pushes the parcel forward along the trajectory and inward with respect to the bend; the Coriolis force pushes inwards (outwards) of the bend in the northern (southern) hemisphere; and friction pulls (necessarily) rearwards.

Governing Equations

For the dynamical equilibrium of the parcel, either component of acceleration times the parcel's mass is equal to the components of the external forces acting in the same direction. As the equations of equilibrium for the parcel are written in natural coordinates, the component equations for the horizontal momentum per unit mass are expressed as follows:

$$\frac{DV}{Dt} = -\frac{1}{\rho}\frac{\partial p}{\partial s} - KV$$

$$\frac{V^2}{R} = -\frac{1}{\rho}\frac{\partial p}{\partial n} \pm fV,$$

In the forward and sideway directions respectively, where ρ is the density of air.

The terms can be broken down as follows:

- DV / Dt is the temporal rate of speed change to the parcel (tangential acceleration);
- $-\partial p / \partial s$ is the component of the pressure force per unit volume along the trajectory;
- $-KV$ is the deceleration due to friction;
- V^2 / R is the centripetal acceleration;

- $-\partial p / \partial n$ is the component of the pressure force per unit volume perpendicular to the trajectory;
- $\pm fV$ is the Coriolis force per unit mass (the sign ambiguity depends on the mutual orientation of the force vector and n).

Steady-State Assumption

In the following discussions, we consider steady-state flow. The speed cannot thus change with time, and the component forces producing *tangential acceleration* need to sum up to zero. In other words, active and resistive forces must balance out in the forward direction in order that $DV / Dt = 0$. Importantly, no assumption is made yet on whether the right-hand side forces are of either significant or negligible magnitude there. Moreover, trajectories and streamlines coincide in steady-state conditions, and the pairs of adjectives tangential/normal and streamwise/cross-stream become interchangeable. An atmospheric flow in which the tangential acceleration is not negligible is called allisobaric.

The velocity direction can still change in space along the trajectory that, excluding inertial flows, is set by the pressure pattern.

General Framework

The schematizations: Omitting specific terms in the tangential and normal balance equations, we obtain one of the five following idealized flows: antitriptic, geostrophic, cyclostrophic, inertial, and gradient flows. By reasoning on the balance of the remaining terms, we can understand

- what arrangement of the pressure field supports such flows;
- along which trajectory the parcel of air travels; and
- with which speed it does so.

The following yes/no table shows which contributions are considered in each idealisation. The Ekman layer's schematisation is also mentioned for completeness, and is treated separately since it involves the internal friction of air rather than that between air and ground.

	Antitriptic flow	*Geostrophic flow*	*Cyclostrophic flow*	*Inertial flow*	*Gradient flow*	*Ekman flow*
curvature	N	N	Y	Y	Y	N
friction	Y	N	N	N	N	Y
pressure	Y	Y	Y	N	Y	Y
Coriolis	N	Y	N	Y	Y	Y

The Limitations

Vertical Differences of Air Properties: The equations were said to apply to parcels of air moving on horizontal planes. Indeed, when one considers a column of atmosphere, it is seldom the case that the air density is the same all the height up, since temperature and moisture content, hence density, do change with height. Every parcel within such a column moves according to the air properties at its own height.

Homogeneous sheets of air may slide one over the other, so long as stable stratification of lighter air on top of heavier air leads to well-separated layers. If some air happens to be heavier/lighter than that in the surroundings, though, vertical motions do occur and modify the horizontal motion in turn. In nature downdrafts and updrafts can sometimes be more rapid and intense than the motion parallel to the ground. The balanced-flow equations do not contain either a force representing the sinking/buoyancy action or the vertical component of velocity.

Consider also that the pressure is normally known through instruments (barometres) near the ground/sea level. The isobars of the ordinary weather charts summarise these pressure measurements, adjusted to the mean sea level for uniformity of presentation, at one particular time. Such values represent the weight of the air column overhead without indicating the details of the changes of the air's specific weight overhead. Also, by Bernoulli's theorem, the measured pressure is not exactly the weight of the air column, should significant vertical motion of air occur. Thus, the pressure force acting on individual parcels of air at different heights is not really known through the measured values. When using information from a surface-pressure chart in balanced-flow formulations, the forces are best viewed as applied to the entire air column.

One difference of air speed in every air column invariably occurs, however, near the ground/sea, also if the air density is the same anywhere and no vertical motion occurs. There, the roughness of the contact surface slows down the air motion above, and this retarding effect peters out with height. See, for example, planetary boundary layer. Frictional antitriptic flow applies near the ground, while the other schematisations apply far enough from the ground not to feel its 'braking' effect (*free-air flow*). This is a reason to keep the two groups conceptually separated. The transition from low-quote to high quote schematisations is bridged by Ekman-like schematisations where air-to-air friction, Coriolis and pressure forces are in balance.

In summary, the balanced-flow speeds apply well to air columns that can be regarded as homogeneous (constant density, no vertical motion) or, at most, stably stratified (non-constant density, yet no vertical motion). An uncertainty in the estimate arises if we are not able to verify these conditions to occur. They also cannot describe the motion of the entire column from the contact surface with the Earth up to the outer atmosphere, because of the on-off handling of the friction forces.

Horizontal Differences of Air Properties

Even if air columns are homogeneous with height, the density of each column can change from location to location, firstly since air masses have different temperatures and moisture content depending on their origin; and then since air masses modify their properties as they flow over Earth's surface. For example, in extra-tropical cyclones the air circulating around a pressure low typically comes with a sector of warmer temperature wedged within colder air. The gradient-flow model of cyclonic circulation does not allow for these features.

Balanced-flow schematisations can be used to estimate the wind speed in air flows covering several degrees of latitude of Earth's surface. However, in this case assuming a constant Coriolis parameter is unrealistic, and the balanced-flow speed can be applied locally.

Unsteadiness

The balanced-flow approach identifies typical trajectories and steady-state wind speeds derived from balance-giving pressure patterns. In reality, pressure patterns and the motion of air masses are tied together, since accumulation (or density increase) of air mass somewhere increases the pressure on the ground and vice versa. Any new pressure gradient will cause a new displacement of air, and thus a continuous rearrangement. As weather itself demonstrates, steady-state conditions are exceptional.

Since friction, pressure gradient and Coriolis forces do not necessarily balance out, air masses actually accelerate and decelerate, so the actual speed depends on its past values too. As seen next, the neat arrangement of pressure fields and flow trajectories, either parallel or at a right angle, in balanced-flow follows from the assumption of steady flow.

The steady-state balanced-flow equations do not explain how the flow was set in motion in the first place. Also, if pressure patterns change quickly enough, balanced-flow speeds cannot help track the air parcels over long distances, simply because the forces that the parcel feels have changed while it is displaced. The particle will end up somewhere else compared to the case that it had followed the original pressure pattern.

In summary, the balanced-flow equations give out consistent steady-state wind speeds that can estimate the situation at a certain moment and a certain place. These speeds cannot be confidently used to understand where the air is moving to in the long run, because the forcing naturally changes or the trajectories are skewed with respect to the pressure pattern.

Antitriptic Flow

Antitriptic flow describes a steady-state flow in a spatially varying pressure field when

- the entire pressure gradient exactly balances friction alone; and:
- all actions promoting curvature are neglected.

The name comes from the Greek words 'anti' (against, counter) and 'triptein' (to rub) - meaning that this kind of flow proceeds by countering friction.

Formulation

In the streamwise momentum equation, friction balances the pressure gradient component without being negligible (so that $K \neq 0$). The pressure gradient vector is only made by the component along the trajectory tangent s. The balance in the streamwise direction determines the antitriptic speed as:

$$V = -\frac{1}{K\rho}\frac{\partial p}{\partial s}$$

A positive speed is guaranteed by the fact that antitriptic flows move along the downward slope of the pressure field, so that mathematically $\partial p / \partial s < 0$. Provided the product KV is constant and ρ stays the same, p turns out to vary linearly with s and the trajectory is such that the parcel feels equal pressure drops while it covers equal distances. (This changes, of course, when using a non-linear model of friction or a coefficient of friction that varies in space to allow for different surface roughness.)

In the cross-stream momentum equation, the Coriolis force and normal pressure gradient are both negligible, leading to no net bending action. As the centrifugal term V^2 / R vanishes while the speed is non-zero, the radius of curvature goes to infinity, and the trajectory must be a straight line. In addition, the trajectory is perpendicular to the isobars since $\partial p / \partial n = 0$. Since this condition occurs when the n direction is that of an isobar, s is perpendicular to the isobars. Thus, antitriptic isobars need to be equispaced circles or straight lines.

Application

Antitriptic flow is probably the least used of the five balanced-flow idealizations, because the conditions are quite strict. However, it is the only one for which the friction underneath is regarded as a primary contribution. Therefore, the antitriptic schematisation applies to flows that take place near the Earth's surface, in a region known as constant-stress layer.

In reality, the flow in the constant-stress layer has a component parallel to the isobars too, since it is often driven by the faster flow overhead. This occurs owing to the so-called *free-air flow* at high quotes, which tends to be parallel to the isobars, and to the Ekman flow at intermediate quotes, which causes a reduction of the free-air speed and a turning of direction while approaching the surface.

Because the Coriolis effects are neglected, antitriptic flow occurs either near the equator (irrespective of the motion's lengthscale) or elsewhere whenever the Ekman number of the flow is large (normally for small-scale processes), as opposed to geostrophic flows.

Antitriptic flow can be used to describe some boundary-layer phenomena such as sea breezes, Ekman pumping, and the low level jet of the Great Plains.

Geostrophic Flow

Geostrophic flow describes a steady-state flow in a spatially varying pressure field when

- frictional effects are neglected; and:
- the entire pressure gradient exactly balances the Coriolis force alone (resulting in no curvature).

The name 'geostrophic' stems from the Greek words 'ge' (Earth) and 'strephein' (to turn). This etymology does not suggest turning of trajectories, rather a rotation around the Earth.

Formulation

In the streamwise momentum equation, negligible friction is expressed by K=0 and, for steady-state balance, negligible streamwise pressure force follows.

The speed cannot be determined by this balance. However, $\partial p / \partial s = 0$ entails that the trajectory must run along isobars, else the moving parcel would experience changes of pressure like in antitriptic flows. No bending is thus only possible if the isobars are straight lines in the first instance. So, geostrophic flows take the appearance of a stream channelled along such isobars.

In the cross-stream momentum equation, non-negligible Coriolis force is balanced by the pressure force, in a way that the parcel does not experience any bending action. Since the trajectory does not bend, the positive orientation of n cannot be determined for lack of a centre of curvature. The signs of the normal vector components become uncertain in this case. However, the pressure force must exactly counterbalance the Coriolis force anyway, so the parcel of air needs to travel with the Coriolis force contrary to the decreasing sideways slope of pressure. Therefore, irrespective of the uncertainty in formally setting the unit vector n, the parcel always travels with the lower pressure at its left (right) in the northern (southern) hemisphere.

The geostrophic speed is

$$V = \frac{1}{\rho}\left|\frac{1}{f}\frac{\partial p}{\partial n}\right|.$$

The expression of geostrophic speed resembles that of antitriptic speed: here the speed is determined by the magnitude of the pressure gradient across (instead of along) the trajectory that develops along (instead of across) an isobar.

Application

Modelers, theoreticians, and operational forecasters frequently make use of geostrophic/quasi-geostrophic approximation. Because friction is unimportant, the geostrophic balance fits flows high enough above the Earth's surface. Because the Coriolis force is relevant, it normally fits processes with small Rossby number, typically having large lengthscales. Geostrophic conditions are also realised for flows having small Ekman number, as opposed to antitriptic conditions.

It is frequent that the geostrophic conditions develop between a well-defined pair of pressure high and low; or that a major geostrophic stream is flanked by several higher- and lower-pressure regions at either side of it. Although the balanced-flow equations do not allow for internal (air-to-air) friction, the flow directions in geostrophic streams and nearby rotating systems are also consistent with shear contact between those.

The speed of a geostrophic stream is larger (smaller) than that in the curved flow around a pressure low (high) with the same pressure gradient: this feature is explained by the more general gradient-flow schematisation. This helps use the geostrophic speed as a back-of-the-envelope estimate of more complex arrangements.

The etymology and the pressure charts shown suggest that geostrophic flows may describe atmospheric motion at rather large scales, although not necessarily so.

Cyclostrophic Flow

Cyclostrophic flow describes a steady-state flow in a spatially varying pressure field when

- the frictional and Coriolis actions are neglected; and:
- the centripetal acceleration is entirely sustained by the pressure gradient.

Trajectories do bend. The name 'cyclostrophic' stems from the Greek words 'kyklos' (circle) and 'strephein' (to turn).

Formulation

Like in geostrophic balance, the flow is frictionless and, for steady-state motion, the trajectories follow the isobars.

In the cross-stream momentum equation, only the Coriolis force is discarded, so that the centripetal acceleration is just the cross-stream pressure force per unit mass

$$\frac{V^2}{R} = -\frac{1}{\rho}\frac{\partial p}{\partial n}.$$

This implies that the trajectory is subject to a bending action, and that the cyclostrophic speed is

$$V = \sqrt{-\frac{R}{\rho}\frac{\partial p}{\partial n}}.$$

So, the cyclostrophic speed is determined by the magnitude of the pressure gradient across the trajectory and by the radius of curvature of the isobar. The flow is faster, the farther away from its centre of curvature, albeit less than linearly.

Another implication of the cross-stream momentum equation is that a cyclostrophic flow can only develop next to a low-pressure area. This is implied in the requirement that the quantity under the square root is positive. Recall that the cyclostrophic trajectory was found to be an isobar. Only if the pressure increases from the centre of curvature outwards, the pressure derivative is negative and the square root is well defined - the pressure in the centre of curvature must thus be a low. The above mathematics gives no clue whether the cyclostrophic rotation ends up to be clockwise or anticlockwise, meaning that the eventual arrangement is a consequence of effects not allowed for in the relationship, namely the rotation of the parent cell.

Application

The cyclostrophic schematisation is realistic when Coriolis and frictional forces are both negligible, that is for flows having large Rossby number and small Ekman number. Coriolis effects are ordinarily negligible in lower latitudes or on smaller scales. Cyclostrophic balance can be achieved in systems such as tornadoes, dust devils and waterspouts. Cyclostrophic speed can also be seen as one of the contribution of the gradient balance-speed, as shown next. Among the studies using the cyclostrophic schematisation, Rennó and Bluestein use the cyclostrophic speed equation to construct a theory for waterspouts; and Winn, Hunyady, and Aulich use the cyclostrophic approximation to compute the maximum tangential winds of a large tornado which passed near Allison, Texas on 8 June 1995.

Inertial Flow

Unlike all other flows, inertial balance implies a uniform pressure field. In this idealisation:

- the flow is frictionless;
- no pressure gradient (and force) is present at all.

The only remaining action is the Coriolis force, which imparts curvature to the trajectory.

Formulation

As before, frictionless flow in steady-state conditions implies that $\partial p / \partial s = 0$. However, in this case isobars are not defined in the first place. We cannot draw any anticipation about the trajectory from the arrangement of the pressure field.

In the cross-stream momentum equation, after omitting the pressure force, the centripetal acceleration is the Coriolis force per unit mass. The sign ambiguity disappears, because the bending is solely determined by the Coriolis force that sets unchallenged the side of curvature - so this force has always a positive sign. The inertial rotation will be clockwise (anticlockwise) in the northern (southern) hemisphere. The momentum equation

$$\frac{V^2}{R} = |f|V,$$

gives us the inertial speed

$$V = |f|R.$$

The inertial speed's equation only helps determine either the speed or the radius of curvature once the other is given. The trajectory resulting from this motion is also known as inertial circle. The balance-flow model gives no clue on the initial speed of an inertial circle, which needs to be triggered by some external perturbation.

Application

Since atmospheric motion is due largely to pressure differences, inertial flow is not very applicable in atmospheric dynamics. However, the inertial speed appears as a contribution to the solution of the gradient speed. Moreover, inertial flows are observed in the ocean streams, where flows are less driven by pressure differences than in air because of higher density—inertial balance can occur at depths such that the friction transmitted by the surface winds downwards vanishes.

Gradient Flow

Gradient flow is an extension of geostrophic flow as it accounts for curvature too, making this a more accurate approximation for the flow in the upper atmosphere. However, mathematically gradient flow is slightly more complex, and geostrophic flow may be fairly accurate, so the gradient approximation is not as frequently mentioned.

Gradient flow is also an extension of the cyclostrophic balance, as it allows for the effect of the Coriolis force, making it suitable for flows with any Rossby number. Finally, it is an extension of inertial balance, as it allows for a pressure force to drive the flow.

Formulation

Like in all but the antitriptic balance, frictional and pressure forces are neglected in the streamwise momentum equation, so that it follows from $\partial p / \partial s = 0$ that the flow is parallel to the isobars.

Solving the full cross-stream momentum equation as a quadratic equation for V yields

$$V = \pm\frac{fR}{2} \pm \sqrt{\frac{f^2R^2}{4} - \frac{R}{\rho}\frac{\partial p}{\partial n}}.$$

Not all solutions of the gradient wind speed yield physically plausible results: the right-hand side as a whole needs be positive because of the definition of speed; and the quantity under square root needs to be non-negative. The first sign ambiguity follows from the mutual orientation of the Coriolis force and unit vector n, whereas the second follows from the square root.

The important cases of cyclonic and anticyclonic circulations are discussed next.

Pressure Lows and Cyclones

For regular cyclones (air circulation around pressure lows), the pressure force is inward (positive term) and the Coriolis force outward (negative term) irrespective of the hemisphere. The cross-trajectory momentum equation is

$$\frac{V^2}{R} = \frac{1}{\rho}\left|\frac{\partial p}{\partial n}\right| - |f|V .$$

Dividing both sides by $|f|\,V$, one recognises that

$$\frac{V_{geostrophic}}{V_{cyclone}} = 1 + \frac{V_{cyclone}}{V_{inertial}} > 1 ,$$

whereby the cyclonic gradient speed V is smaller than the corresponding geostrophic, less accurate estimate, and naturally approaches it as the radius of curvature grows (as the inertial velocity goes to infinity). In cyclones, therefore, curvature slows down the flow compared to the no-curvature value of geostrophic speed.

The positive root of the cyclone equation is

$$V_{cyclone} = -\frac{V_{inertial}}{2} + \sqrt{\frac{V_{inertial}^2}{4} + V_{cyclostrophic}^2} .$$

This speed is always well defined as the quantity under the square root is always positive.

Pressure Highs and Anticyclones

In anticyclones (air circulation around pressure highs), the Coriolis force is always inward (and positive), and the pressure force outward (and negative) irrespective of the hemisphere. The cross-trajectory momentum equation is

$$\frac{V^2}{R} = -\frac{1}{\rho}\left|\frac{\partial p}{\partial n}\right| + |f|V .$$

Dividing both sides by $|f|\,V$, we obtain

$$\frac{V_{geostrophic}}{V_{anticyclone}} = 1 - \frac{V_{anticyclone}}{V_{inertial}} < 1,$$

whereby the anticyclonic gradient speed V is larger than the geostrophic value and approaches it as the radius of curvature becomes larger. In anticyclones, therefore, the curvature of isobars speeds up the airflow compared to the (geostrophic) no-curvature value.

There are two positive roots for V, but the only one consistent with the limit to geostrophic conditions is

$$V_{anticyclone} = \frac{V_{inertial}}{2} - \sqrt{\frac{V_{inertial}^2}{4} - V_{cyclostrophic}^2}$$

that requires that $V_{inertial} \geq 2V_{cyclostrophic}$ to be meaningful. This condition can be translated in the requirement that, given a high-pressure zone with a constant pressure slope at a certain latitude, there must be a circular region around the high without wind. On its circumference the air blows at half the corresponding inertial speed (at the cyclostrophic speed), and the radius is

$$R^* = \frac{4}{\rho f^2}\left|\frac{\partial p}{\partial n}\right|,$$

obtained by solving the above inequality for R. Outside this circle the speed decreases to the geostrophic value as the radius of curvature increases. The width of this radius grows with the intensity of the pressure gradient.

Application

Gradient Flow is useful in studying atmospheric flow rotating around high and low pressures centres with small Rossby numbers. This is the case where the radius of curvature of the flow about the pressure centres is small, and geostrophic flow no longer applies with a useful degree of accuracy.

Balanced-Flow Speeds Compared.

Each balanced-flow idealisation gives a different estimate for the wind speed in the same conditions. Here we focus on the schematisations valid in the upper atmosphere. Firstly, imagine that a sample parcel of air flows 500 metres above the sea surface, so that frictional effects are already negligible. The density of (dry) air at 500 metre above the mean sea level is 1.167 kg/m^3 according to its equation of state.

Secondly, let the pressure force driving the flow be measured by a rate of change taken as 1hPa/100 km (an average value). Recall that it is not the value of the pressure to be important, but the slope with which it changes across the trajectory. This slope applies equally well to the spacing of straight isobars (geostrophic flow) or of curved isobars (cyclostrophic and gradient flows).

Thirdly, let the parcel travel at a latitude of 45 degrees, either in the southern or northern hemisphere—so the Coriolis force is at play with a Coriolis parameter of 0.000115 Hz.

The balance-flow speeds also changes with the radius of curvature R of the trajectory/isobar. In case of circular isobars, like in schematic cyclones and anticyclones, the radius of curvature is also the distance from the pressure low and high respectively.

Taking two of such distances R as 100 km and 300 km, the speeds are (in m/s)

	Geostrophic	***Cyclostrophic***	***Inertial***	***Gradient (H-pressure)***	***Gradient (L-pressure)***
R=100 km	7.45	9.25	11.50	N/A	5.15
R=300 km	7.45	16.00	34.50	10.90	6.30

The chart shows how the different speeds change in the conditions chosen above and with increasing radius of curvature.

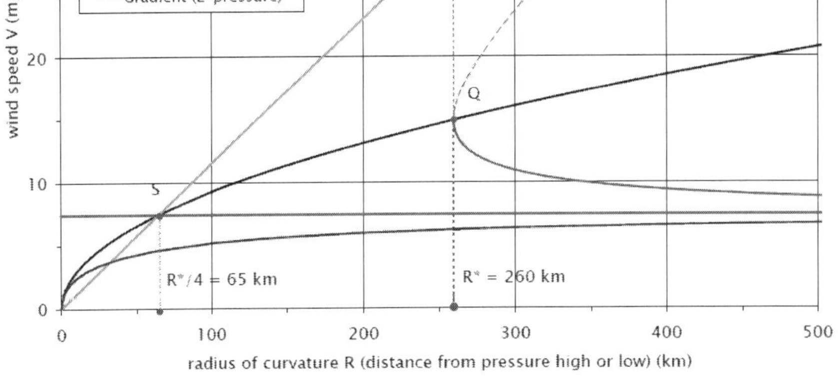

The geostrophic speed (pink line) does not depend on curvature at all, and it appears as a horizontal line. However, the cyclonic and anticyclonic gradient speeds approach it as the radius of curvature becomes indefinitely large—geostrophic balance is indeed the limiting case of gradient flow for vanishing centripetal acceleration (that is, for pressure and Coriolis force exactly balancing out).

The cyclostrophic speed (black line) increases from zero and its rate of growth with R is less than linear. In reality an unbounded speed growth is impossible because the conditions supporting the flow change at some distance. Also recall that the cyclostrophic conditions apply to small-scale processes, so extrapolation to higher radii is physically meaningless.

The inertial speed (green line), which is independent of the pressure gradient that we chose, increases linearly from zero and it soon becomes much larger than any other.

The gradient speed comes with two curves valid for the speeds around a pressure low (blue) and a pressure high (red). The wind speed in cyclonic circulation grows from zero as the radius increases and is always less than the goestrophic estimate.

In the anticyclonic-circulation example, there is no wind within the distance of 260 km (point R*) — this is the area of no/low winds around a pressure high. At that distance the first anticyclonic wind has the same speed as the cyclostrophic winds (point Q), and half of that of the inertial wind (point P). Farther away from point R*, the anticyclonic wind slows down and approaches the geostrophic value with decreasingly larger speeds.

There is also another noteworthy point in the curve, labelled as S, where inertial, cyclostrophic and geostrophic speeds are equal. The radius at S is always a fourth of R*, that is 65 km here.

Some limitations of the schematisations become also apparent. For example, as the radius of curvature increases along a meridian, the corresponding change of latitude implies different values of the Coriolis parameter and, in turn, force. Conversely, the Coriolis force stays the same if the radius is along a parallel. So, in the case of circular flow, it is unlikely that the speed of the parcel does not change in time around the full circle, because the air parcel will feel the different intensity of the Coriolis force as it travels across different latitudes. Additionally, the pressure fields quite rarely take the shape of neat circular isobars that keep the same spacing all around the circle. Also, important differences of density occur in the horizontal plan as well, for example when warmer air joins the cyclonic circulation, thus creating a warm sector between a cold and a warm front.

Terminology in Fluid Dynamics

The concept of pressure is central to the study of both fluid statics and fluid dynamics. A pressure can be identified for every point in a body of fluid, regardless of whether the fluid is in motion or not. Pressure can be measured using an aneroid, Bourdon tube, mercury column, or various other methods.

Some of the terminology that is necessary in the study of fluid dynamics is not found in other similar areas of study. In particular, some of the terminology used in fluid dynamics is not used in fluid statics.

Terminology in Incompressible Fluid Dynamics

The concepts of total pressure and dynamic pressure arise from Bernoulli's equation and are significant in the study of all fluid flows. (These two pressures are not pressures in the usual sense—they cannot be measured using an aneroid, Bourdon tube or mercury column.) To avoid potential ambiguity when referring to pressure in fluid dynamics, many authors use the term static pressure to distinguish it from total pressure and dynamic pressure. Static pressure is identical to pressure and can be identified for every point in a fluid flow field.

In *Aerodynamics*, L.J. Clancy writes: *To distinguish it from the total and dynamic pressures, the actual pressure of the fluid, which is associated not with its motion but with its state, is often referred to as the static pressure, but where the term pressure alone is used it refers to this static pressure.*

A point in a fluid flow where the flow has come to rest (i.e. speed is equal to zero adjacent to some solid body immersed in the fluid flow) is of special significance. It is of such importance that it is given a special name—a stagnation point. The static pressure at the stagnation point is of special significance and is given its own name—stagnation pressure. In incompressible flows, the stagnation pressure at a stagnation point is equal to the total pressure throughout the flow field.

Terminology in compressible fluid dynamics

In a compressible fluid, such as air, the temperature and density are essential when determining the state of the fluid. In addition to the concept of total pressure (also known as stagnation pressure), the concepts of total (or stagnation) temperature and total (or stagnation) density are also essential in any study of compressible fluid flows. To avoid potential ambiguity when referring to temperature and density, many authors use the terms static temperature and static density. Static temperature is identical to temperature; and static density is identical to density; and both can be identified for every point in a fluid flow field.

The temperature and density at a stagnation point are called stagnation temperature and stagnation density.

A similar approach is also taken with the thermodynamic properties of compressible fluids. Many authors use the terms total (or stagnation) enthalpy and total (or stagnation) entropy. The terms static enthalpy and static entropy appear to be less common, but where they are used they mean nothing more than enthalpy and entropy respectively, and the prefix "static" is being used to avoid ambiguity with their 'total' or 'stagnation' counterparts. Because the 'total' flow conditions are defined by isentropically bringing the fluid to rest, the total (or stagnation) entropy is by definition always equal to the "static" entropy.

Fluid

In physics, a fluid is a substance that continually deforms (flows) under an applied shear stress. Fluids are a subset of the phases of matter and include liquids, gases, plasmas and, to some extent, plastic solids.

Although the term "fluid" includes both the liquid and gas phases, in common usage, "fluid" is often used as a synonym for "liquid", with no implication that gas could also be present. For example, "brake fluid" is hydraulic oil and will not perform its required function if there is gas in it. This colloquial usage of the term is also common in medicine and in nutrition ("take plenty of fluids").

Liquids form a free surface (that is, a surface not created by the container) while gases do not. The distinction between solids and fluid is not entirely obvious. The distinction is made by evaluating the viscosity of the substance. Silly Putty can be considered to behave like a solid or a fluid, depending on the time period over which it is observed. It is best described as a viscoelastic fluid. There are many examples of substances proving difficult to classify. A particularly interesting one is pitch, as demonstrated in the pitch drop experiment currently running at the University of Queensland.

Physics

Fluids display properties such as:

- not resisting deformation, or resisting it only lightly (viscosity), and
- the ability to flow (also described as the ability to take on the shape of the container). This also means that all fluids have the property of fluidity.

These properties are typically a function of their inability to support a shear stress in static equilibrium.

Solids can be subjected to shear stresses, and to normal stresses—both compressive and tensile. In contrast, ideal fluids can only be subjected to normal, compressive stress which is called pressure. Real fluids display viscosity and so are capable of being subjected to low levels of shear stress.

Modelling

In a solid, shear stress is a function of strain, but in a fluid, shear stress is a function of strain rate. A consequence of this behaviour is Pascal's law which describes the role of pressure in characterizing a fluid's state.

Depending on the relationship between shear stress, and the rate of strain and its derivatives, fluids can be characterized as one of the following:

- Newtonian fluids : where stress is directly proportional to rate of strain
- Non-Newtonian fluids : where stress is not proportional to rate of strain, its higher powers and derivatives.

The behaviour of fluids can be described by the Navier–Stokes equations—a set of partial differential equations which are based on:

- continuity (conservation of mass),
- conservation of linear momentum,
- conservation of angular momentum,
- conservation of energy.

The study of fluids is fluid mechanics, which is subdivided into fluid dynamics and fluid statics depending on whether the fluid is in motion.

Chapter 3

Pipe Flow

Pipe flow, a branch of Hydraulics and Fluid Mechanics, is a type of liquid flow within a closed conduit (conduit in the sense of a means of containment). The other type of flow within a conduit is open channel flow. These two types of flow are similar in many ways, but differ in one important respect. Pipe flow does not have a free surface which is found in open-channel flow. Pipe flow, being confined within closed conduit, does not exert direct atmospheric pressure, but does exert hydraulic pressure on the conduit.

Not all flow within a closed conduit is considered pipe flow. Storm sewers are closed conduits but usually maintain a free surface and therefore are considered open-channel flow. The exception to this is when a storm sewer operates at full capacity, and then can become pipe flow. Energy in pipe flow is expressed as head and is defined by the Bernoulli equation. In order to conceptualize head along the course of flow within a pipe, diagrams often contain a hydraulic grade line. Pipe flow is subject to frictional losses as defined by the Darcy-Weisbach formula.

States of Flow

The behaviour of pipe flow is governed mainly by the effects of viscosity and gravity relative to the inertial forces of the flow. Depending on the effect of viscosity relative to inertia, as represented by the Reynolds number, the flow can be either laminar or turbulent. At a Reynolds number below the critical value of approximately 2040 pipe flow will ultimately be laminar, whereas above the critical value turbulent flow can persist. In addition, the transition between laminar flow and turbulence can be sensitive to disturbance levels and imperfections.

Flow through pipes can roughly be divided into two:

- Laminar flow
- Turbulent flow

Hagen–Poiseuille Equation

In fluid dynamics, the Hagen–Poiseuille equation, also known as the Hagen–Poiseuille law, Poiseuille law or Poiseuille equation, is a physical law that gives the pressure drop in a fluid flowing through a long cylindrical pipe. It can be successfully applied to air flow in lung alveoli, for the flow through a drinking straw or through a hypodermic needle. It was experimentally derived independently by Gotthilf Heinrich Ludwig Hagen in 1839 and Jean Louis Marie Poiseuille in 1838, and published by Poiseuille in 1840 and 1846.

The assumptions of the equation are that the fluid is incompressible and newtonian; the flow is laminar through a pipe of constant circular cross-section that is substantially longer than its diameter; and there is no acceleration of fluid in the pipe. For velocities and pipe diameters above a threshold, actual fluid flow is not laminar but turbulent, leading to larger pressure drops than calculated by the Hagen–Poiseuille equation.

Equation

Standard fluid dynamics notation:

In standard fluid dynamics notation:

$$\Delta P = \frac{8\mu L Q}{\pi r^4}$$

or

$$\Delta P = \frac{128\mu L Q}{\pi d^4}$$

where:

ΔP is the pressure loss

L is the length of pipe

μ is the dynamic viscosity

Q is the volumetric flow rate

r is the radius

d is the diameter

π is the mathematical constant Pi

Physics Notation

$$\Phi = \frac{dV}{dt} = v\pi R^2 = \frac{\pi R^4}{8\eta}\left(\frac{-\Delta P}{\Delta x}\right) = \frac{\pi R^4}{8\eta}\frac{|\Delta P|}{L}$$

where in compatible units (e.g., S.I.):

Φ is the volumetric flow rate (denoted as Q above)

$V(t)$ is the volume of the liquid transferred as a function of time,

v is mean fluid velocity along the length of the tube

x is distance in direction of flow

R is the internal radius of the tube

ΔP is the pressure difference between the two ends

is the dynamic fluid viscosity (S.I. unit: pascal-second (Pa·s)),

L is the length of the tube

The equation does not hold close to the pipe entrance.

The equation fails in the limit of low viscosity, wide and/or short pipe. Low viscosity or a wide pipe may result in turbulent flow, making it necessary to use more complex models, such as Darcy-Weisbach equation. If the pipe is too short, the Hagen–Poiseuille equation may result in unphysically high flow rates; the flow is bounded by Bernoulli's principle, under less restrictive conditions, by

$$\Phi_{max} = \pi R^2 \sqrt{2\Delta P / \rho}\,.$$

Relation to Darcy–Weisbach

This result is also a solution to the phenomenological Darcy–Weisbach equation in the field of hydraulics, given a relationship for the friction factor in terms of the Reynolds number:

$$\Lambda = \frac{64}{\mathrm{Re}}, \qquad \mathrm{Re} = \frac{2\rho v r}{\eta},$$

where *Re* is the Reynolds number and ρ fluid density. In this form the law approximates the *Darcy friction factor*, the *energy (head) loss factor*, *friction loss factor* or *Darcy (friction) factor* Ë in the laminar flow at very low velocities in cylindrical tube. The theoretical derivation of a slightly different form of the law was made independently by Wiedman in 1856 and Neumann and E. Hagenbach in 1858 (1859, 1860). Hagenbach was the first who called this law the Poiseuille's law.

The law is also very important specially in hemorheology and hemodynamics, both fields of physiology.

The Poiseuilles' law was later in 1891 extended to turbulent flow by L. R. Wilberforce, based on Hagenbach's work.

Liquid Flow through a Pipe

Assume the liquid exhibits Laminar flow. Laminar flow in a round pipe prescribes that there are a bunch of circular layers (lamina) of liquid, each having a velocity determined only by their radial distance from the centre of the tube. Also assume the centre is moving fastest while the liquid touching the walls of the tube is stationary (due to the no-slip condition).

To figure out the motion of the liquid, all forces acting on each lamina must be known:

1. The pressure force pushing the liquid through the tube is the change in pressure multiplied by the area: $F = -\Delta PA$. This force is in the direction of the motion of the liquid. The negative sign comes from the conventional way we define $\Delta P = P_{end} - P_{top} < 0$.
2. Viscosity effects will pull from the faster lamina immediately closer to the centre of the tube.
3. Viscosity effects will drag from the slower lamina immediately closer to the walls of the tube.

Viscosity

Figure: *Two fluids moving past each other in the* x *direction. The liquid on top is moving faster and will be pulled in the negative direction by the bottom liquid while the bottom liquid will be pulled in the positive direction by the top liquid.*

When two layers of liquid in contact with each other move at different speeds, there will be a shear force between them. This force is proportional to the area of contact A, the velocity gradient in the direction of flow $\Delta v_x / \Delta y$, and a proportionality constant η (viscosity) and is given by

$$F_{\text{viscosity, top}} = -\eta A \frac{\Delta v_x}{\Delta y}.$$

The negative sign is in there because we are concerned with the faster moving liquid (top in figure), which is being slowed by the slower liquid (bottom in figure). By Newton's third law of motion, the force on the slower liquid is equal and opposite (no negative sign) to the force on the faster liquid. This equation assumes that the area of contact is so large that we can ignore any effects from the edges and that the fluids behave as Newtonian fluids.

Faster Lamina

Assume that we are figuring out the force on the lamina with radius *s*. From the equation above, we need to know the area of contact and the velocity gradient. Think of the lamina as a ring of radius *r*, thickness *dr*, and length Δx. The area of contact between the lamina and the faster one is simply the area of the inside of the cylinder: $A = 2\pi r \Delta x$. We don't know the exact form for the velocity of the liquid within the tube yet, but we do know (from our assumption above) that it is dependent on the radius. Therefore, the velocity gradient is the change of the velocity with respect to the change in the radius at the intersection of these two laminae. That intersection is at a radius of *r*. So, considering that this force will be positive with respect to the movement of the liquid (but the derivative of the velocity is negative), the final form of the equation becomes

$$F_{\text{viscosity, fast}} = -\eta 2\pi r \Delta x \left.\frac{dv}{dr}\right|_r$$

where the vertical bar and subscript *r* following the derivative indicates that it should be taken at a radius of *r*.

Slower Lamina

Next let's find the force of drag from the slower lamina. We need to calculate the same values that we did for the force from the faster lamina. In this case, the area of contact is at *r+dr* instead of *r*. Also, we need to remember that this force opposes the direction of movement of the liquid and will therefore be negative (and that the derivative of the velocity is negative).

$$F_{\text{viscosity, slow}} = \eta 2\pi (r + dr) \Delta x \left.\frac{dv}{dr}\right|_{r+dr}$$

Putting it All Together

To find the solution for the flow of liquid through a tube, we need to make one last assumption. There is no acceleration of liquid in the pipe, and by Newton's first law, there is no net force. If there is no net force then we can add all of the forces together to get zero

$$0 = F_{\text{pressure}} + F_{\text{viscosity, fast}} + F_{\text{viscosity, slow}}$$

or

$$0 = -\Delta P 2\pi r dr - \eta 2\pi r \Delta x \left.\frac{dv}{dr}\right|_r + \eta 2\pi (r + dr)\Delta x \left.\frac{dv}{dr}\right|_{r+dr}.$$

First, to get everything happening at the same point, use the first two terms of a Taylor series expansion of the velocity gradient:

$$\left.\frac{dv}{dr}\right|_{r+dr} = \left.\frac{dv}{dr}\right|_r + \left.\frac{d^2v}{dr^2}\right|_r dr.$$

The expression is valid for all laminae. Grouping like terms and dropping the vertical bar since all derivatives are assumed to be at radius *r*,

$$0 = -\Delta P 2\pi r dr + \eta 2\pi dr \Delta x \frac{dv}{dr} + \eta 2\pi r dr \Delta x \frac{d^2v}{dr^2} + \eta 2\pi (dr)^2 \Delta x \frac{d^2v}{dr^2}.$$

Finally, put this expression in the form of a differential equation, dropping the term quadratic in *dr*.

$$\frac{1}{\eta}\frac{\Delta P}{\Delta x} = \frac{d^2v}{dr^2} + \frac{1}{r}\frac{dv}{dr}$$

It can be seen that both sides of the equations are negative: there is a drop of pressure along the tube (left side) and both first and second derivatives of the velocity are negative (velocity has a maximum value at the centre of the tube, where $r = 0$). Using the chain rule, the equation may be re-arranged to:

$$\frac{1}{\eta}\frac{\Delta P}{\Delta x} = \frac{1}{r}\frac{d}{dr} r \frac{dv}{dr}.$$

This differential equation is subject to the following boundary conditions:

$v(r) = 0$ at $r = R$ — "no-slip" boundary condition at the wall

$\frac{dv}{dr} = 0$ at $r = 0$ — axial symmetry.

Axial symmetry means that the velocity v(r) is maximum at the centre of the tube, therefore the first derivative $\frac{dv}{dr}$ is zero at r = 0.

The differential equation can be integrated to:

$$v(r) = \frac{1}{4\eta} r^2 \frac{\Delta P}{\Delta x} + A\ln(r) + B.$$

To find A and B, we use the boundary conditions.

First, the symmetry boundary condition indicates:

$$\frac{dv}{dr} = \frac{1}{2\eta} r \frac{\Delta P}{\Delta x} + A\frac{1}{r} = 0 \text{ at r = 0.}$$

A solution possible only if A = 0. Next the no-slip boundary condition is applied to the remaining equation:

$$v(R) = \frac{1}{4\eta} R^2 \frac{\Delta P}{\Delta x} + B = 0$$

so therefore

$$B = -\frac{1}{4\eta} R^2 \frac{\Delta P}{\Delta x}.$$

Now we have a formula for the velocity of liquid moving through the tube as a function of the distance from the centre of the tube

$$v = -\frac{1}{4\eta}\frac{\Delta P}{\Delta x}(R^2 - r^2)$$

or, at the centre of the tube where the liquid is moving fastest (r = 0) with R being the radius of the tube,

$$v_{max} = -\frac{1}{4\eta}\frac{\Delta P}{\Delta x} R^2.$$

Poiseuille's Law

To get the total volume that flows through the tube, we need to add up the contributions from each lamina. To calculate the flow through each lamina, we multiply the velocity (from above) and the area of the lamina.

$$\Phi(r)dr = \frac{1}{4\eta}\frac{|\Delta P|}{\Delta x}(R^2 - r^2)2\pi r dr = \frac{\pi}{2\eta}\frac{|\Delta P|}{\Delta x}(rR^2 - r^3)dr$$

Finally, we integrate over all lamina via the radius variable r.

$$\Phi = \frac{\pi}{2\eta}\frac{|\Delta P|}{\Delta x}\int_0^R (rR^2 - r^3)dr = \frac{|\Delta P|\pi R^4}{8\eta\Delta x}$$

Poiseuille's Equation for Compressible Fluids

For a compressible fluid in a tube the volumetric flow rate and the linear velocity is not constant along the tube. The flow is usually expressed at outlet pressure. As fluid is compressed or expands, work is done and the fluid is heated and cooled. This means that the flow rate depends on the heat transfer to and from the fluid. For an ideal gas in the isothermal case, where the temperature of the fluid is permitted to equilibrate with its surroundings, and when the pressure difference between ends of the pipe is small, the volumetric flow rate at the pipe outlet is given by

$$\Phi = \frac{dV}{dt} = v\pi R^2 = \frac{\pi R^4 (P_i - P_o)}{8\eta L} \times \frac{P_i + P_o}{2P_o} = \frac{\pi R^4}{16\eta L}\left(\frac{P_i^2 - P_o^2}{P_o}\right)$$

where:

P_i inlet pressure

P_o outlet pressure

$L$is the length of tube

η is the viscosity

R is the radius

V is the volume of the fluid at outlet pressure

v is the velocity of the fluid at outlet pressure

This is usually a good approximation when the flow velocity is less than mach 0.3

This equation can be seen as Poiseuille's law with an extra correction factor $\frac{P_i + P_o}{2} \times \frac{1}{P_o}$ expressing the average pressure relative to the outlet pressure.

Electrical Circuits Analogy

Electricity was originally understood to be a kind of fluid. This hydraulic analogy is still conceptually useful for understanding circuits. This analogy is also used to study the frequency response of fluid mechanical networks using circuit tools, in which case the fluid network is termed a hydraulic circuit.

Poiseuille's law corresponds to Ohm's law for electrical circuits ($V = IR$), where the pressure drop ΔP is analogous to the voltage V and volumetric flow rate Φ is analogous to the current I. Then the resistance

$$R = \frac{8\eta\Delta x}{\pi r^4}.$$

This concept is useful because the effective resistance in a tube is inversely proportional to the fourth power of the radius. This means that halving the radius of the tube increases the resistance to fluid movement by a factor of 16. Both Ohm's law and Poiseuille's law illustrate transport phenomena.

Moody Chart

The Moody chart or Moody diagram is a graph in non-dimensional form that relates the Darcy-Weisbach friction factor, Reynolds number and relative roughness for fully developed flow in a circular pipe. It can be used for working out pressure drop or flow rate down such a pipe.

Description

This dimensionless chart is used to work out pressure drop, ΔP (Pa) (or head loss, h_f (m)) and flow rate through pipes. Head loss can be calculated using the Darcy–Weisbach equation:

$$h_f = f\frac{l}{d}\frac{V^2}{2g};$$

(not to be confused with the Fanning equation and the Fanning friction factor:

$$h_f = 4f\frac{l}{d}\frac{V^2}{2g},$$

which uses a friction-factor equal to one fourth the Darcy-Weisbach friction factor).

Pressure drop can then be evaluated as:

$$\Delta P = \rho g h_f \text{ or directly from } \Delta P = f\frac{\rho V^2}{2}\frac{l}{d},$$

where ρ is the density of the fluid, V is the average velocity in the pipe, f is the friction factor from the Moody chart, l is the length of the pipe and d is the pipe diameter.

The basic chart plots Darcy–Weisbach friction factor against Reynolds number for a variety of relative roughnesses and flow regimes. The relative roughness being the ratio of the mean height of roughness of the pipe to the pipe diameter or $\frac{\epsilon}{d}$.

The Moody chart can be divided into two regimes of flow: laminar and turbulent. For the laminar flow regime, the Darcy–Weisbach friction factor was determined analytically by Poiseuille and $\frac{64}{Re}$ is used. In this regime roughness has no discernible effect. For the turbulent flow regime, the relationship between the friction factor and the Reynolds number is more complex and is governed by the Colebrook equation which is implicit in f:

$$\frac{1}{\sqrt{f}} = -2.0\log_{10}\left(\frac{\frac{\epsilon}{d}}{3.7} + \frac{2.51}{Re\sqrt{f}}\right), \text{turbulent flow.}$$

In 1944, Lewis Ferry Moody plotted the Darcy–Weisbach friction factor into what is now known as the Moody chart. The Fanning friction factor is 1/4 the Darcy–Weisbach one and the equation for pressure drop has a compensating factor of four.

Open-Channel Flow

Open-channel flow, a branch of hydraulics, is a type of liquid flow within a conduit with a free surface, known as a channel. The other type of flow within a conduit is pipe flow. These two types of flow are similar in many ways, but differ in one important respect: the free surface. Open-channel flow has a free surface, whereas pipe flow does not.

Classifications of Flow

Open-channel flow can be classified and described in various ways based on the change in flow depth with respect to time and space. The fundamental type of flow dealt with in open-channel hydraulics.

Time as the Criterion

- *Steady flow* – The depth of flow does not change over time, or if it can be assumed to be constant during the time interval under consideration.
- *Unsteady flow* - The depth of flow does change with time.

Space as the Criterion

- *Uniform flow*: The depth of flow is the same at every section of the channel. Uniform flow can be steady or unsteady, depending on whether or not the depth changes with time, (although unsteady uniform flow is rare).

- *Varied flow:* The depth of flow changes along the length of the channel. Varied flow technically may be either steady or unsteady.
- Varied flow can be further classified as either rapidly or gradually varied.

Rapidly varied flow: The depth changes abruptly over a comparatively short distance. Rapidly varied flow is known as local phenomenon. Examples are the hydraulic jump and the hydraulic drop.

Gradually varied flow: The depth changes over a long distance.

Continuous flow: The discharge is constant throughout the reach of the channel under consideration. This is often the case with steady flow. This flow is considered continuous and therefore can be described using the continuity equation for continuous steady flow.

Spatially varied or discontinuous flow - The discharge of a steady flow is nonuniform along a channel. This happens when water enters and/or leaves the channel along the course of flow. An example of flow entering a channel would be a road side gutter. An example of flow leaving a channel would be an irrigation channel. This flow can be described using the continuity equation for continuous unsteady flow requires the consideration of the time effect and includes a time element as a variable.

States of Flow

The behaviour of open channel flow is governed by the effects of viscosity and gravity relative to the inertial forces of the flow. Surface tension has a minor contribution, but does not play a significant enough role in most circumstances to be a governing factor. Depending on the effect of viscosity relative to inertia, as represented by the Reynolds number, the flow can be either laminar, turbulent, or transitional.

Surge Control

Surge control is used to prevent pressure surges, which can occur in a hydraulic system when the flow in a pipeline is stopped too quickly. When a valve closes or a pump stops the water or oil or other fluid in a hydraulic system will attempt to continue flowing. This will cause an increase in pressure which can damage the pipe. Surge control devices absorb and limit this pressure increase and prevent damage to the pipework.

What is Hydraulic Surge

Hydraulic pressure (transient flow) occurs when the flow of fluid in a pipeline is abruptly changed. Piping systems that use quick acting valves, or use pumps that start up or shut down rapidly, are susceptible to pressure transients (surges). These conditions can result in piping failure, damage to pumps, fittings, instrumentation, and other system components.

Controlling Hydraulic Surge

Controlling hydraulic surges can increase the life of hydraulic systems by effectively managing these destructive forces. Fast acting surge relief valves, Surge suppressors, bladder surge tanks, and hydropneumatic device (surge tank) can be used to control surges. Please refer to article "Specifying surge relief valves in liquid pipelines" with Intec Journal July/August 2012 of ISA.

Typical Applications

Surge control products are currently protecting fluid systems in numerous industries and defence industries, including airport refuelling systems, bulk marine loading/unloading facilities, hydrocarbon pipeline system, fire systems (as surge tanks), municipal water and sewage pump systems, petro-chemical facilities, cogeneration plants, mobile fuel trucks, and diesel fuel loading

Surge Relief Valves

Surge Relief Valves are used in most applications including marine loading/unloading application, liquid pipeline transfer systems, fire protection and water pumping stations. They utilise a differential pressure valve to control pressure transients. A surge relief valve for a liquid transporting pipeline, comprising a surge relief line adapted to be tapped from the main transportation pipeline. The relief line includes the surge relief valve which is conditioned to open very quickly to substantial flow when the pressure in the line exceeds a predetermined safe limit, which is well above the normal pipeline pressure.

When the surge relief valve opens very quickly, it diverts to a reservoir called the surge relief tank (Floating roof, slop tank, closed tank etc.), that quantity of liquid which is necessary to prevent the pressure from exceeding said predetermined limit. Hence, in the event of the surge, much of the pressure is absorbed in the liquid and pipe, and just that quantity of liquid which is necessary to relieve pressures of unsafe proportions is discharged to the surge relief tank.

A return line can be employed to evacuate the reservoir to the main pipeline when the surge subsides.

Hydropneumatic Surge Tanks for Water, Waste Water, and Sewage Service

Water, waste water, and sewage pumping applications, incorporate a hydraulically designed nozzle to dampen destructive energy and prevent vortexing.

Hydraulic Machinery

Hydraulic machines are machinery and tools that use liquid fluid power to do simple work. Heavy equipment is a common example.

In this type of machine, hydraulic fluid is transmitted throughout the machine to various hydraulic motors and hydraulic cylinders and which becomes pressurised according to the resistance present. The fluid is controlled directly or automatically by control valves and distributed through hoses and tubes. The popularity of hydraulic machinery is due to the very large amount of power that can be transferred through small tubes and flexible hoses, and the high power density and wide array of actuators that can make use of this power. Hydraulic machinery is operated by the use of hydraulics, where a liquid is the powering medium.

Force and Torque Multiplication

A fundamental feature of hydraulic systems is the ability to apply force or torque multiplication in an easy way, independent of the distance between the input and output, without the need for mechanical gears or levers, either by altering the effective areas in two connected cylinders or the effective displacement (cc/rev) between a pump and motor. In normal cases, hydraulic ratios are combined with a mechanical force or torque ratio for optimum machine designs such as boom movements and trackdrives for an excavator.

Examples

Two Hydraulic Cylinders Interconnected: Cylinder C1 is one inch in radius, and cylinder C2 is ten inches in radius. If the force exerted on C1 is 10 lbf, the force exerted by C2 is 1000 lbf because C2 is a hundred times larger in area ($S = \pi r^2$) as C1. The downside to this is that you have to move C1 a hundred inches to move C2 one inch. The most common use for this is the classical hydraulic jack where a pumping cylinder with a small diameter is connected to the lifting cylinder with a large diameter.

Pump and Motor

If a hydraulic rotary pump with the displacement 10 cc/rev is connected to a hydraulic rotary motor with 100 cc/rev, the shaft torque required to drive the pump is 10 times less than the torque available at the motor shaft, but the shaft speed (rev/min) for the motor is 10 times less than the pump shaft speed. This combination is actually the same type of force multiplication as the cylinder example (1) just that the linear force in this case is a rotary force, defined as torque.

Both these examples are usually referred to as a hydraulic transmission or hydrostatic transmission involving a certain hydraulic "gear ratio".

Hydraulic Circuits

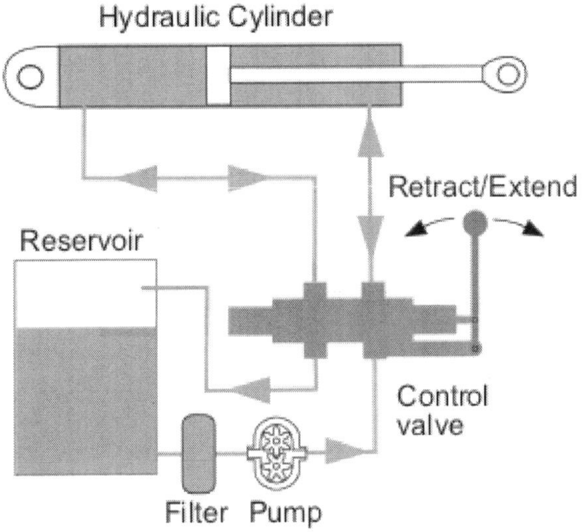

***Figure:** A simple* open centre *hydraulic circuit.*

For the hydraulic fluid to do work, it must flow to the actuator and/or motors, then return to a reservoir. The fluid is then filtered and re-pumped. The path taken by hydraulic fluid is called a hydraulic circuit of which there are several types. Open centre circuits use pumps which supply a continuous flow. The flow is returned to *tank* through the control valve's *open centre*; that is, when the control valve is centred, it provides an open return path to tank and the fluid is not pumped to a high pressure. Otherwise, if the control valve is actuated it routes fluid to and from an actuator and tank. The fluid's pressure will rise to meet any resistance, since the pump has a constant output. If the pressure rises too high, fluid returns to tank through a pressure relief valve. Multiple control valves may be stacked in series [1]. This type of circuit can use inexpensive, constant displacement pumps.

Closed centre circuits supply full pressure to the control valves, whether any valves are actuated or not. The pumps vary their flow rate, pumping very little hydraulic fluid until the operator actuates a valve. The valve's spool therefore doesn't need an open centre return path to tank. Multiple valves can be connected in a parallel arrangement and system pressure is equal for all valves.

Constant Pressure and Load-Sensing Systems

The closed centre circuits exist in two basic configurations, normally related to the regulator for the variable pump that supplies the oil:

Constant pressure systems (CP-system), standard. Pump pressure always equals the pressure setting for the pump regulator. This setting must cover the maximum required load pressure. Pump delivers flow according to required sum of flow to the consumers. The CP-system generates large power losses if the machine works with large variations in load pressure and the average system pressure is much lower than the pressure setting for the pump regulator. CP is simple in design. Works like a pneumatic system. New hydraulic functions can easily be added and the system is quick in response.

Constant pressure systems (CP-system), unloaded. Same basic configuration as 'standard' CP-system but the pump is unloaded to a low stand-by pressure when all valves are in neutral position. Not so fast response as standard CP but pump lifetime is prolonged.

Load-sensing systems (LS-system) generates less power losses as the pump can reduce both flow and pressure to match the load requirements, but requires more tuning than the CP-system with respect to system stability. The LS-system also requires additional logical valves and compensator valves in the directional valves, thus it is technically more complex and more expensive than the CP-system. The LS-system system generates a constant power loss related to the regulating pressure drop for the pump regulator:

$$Powerloss = \Delta p_{LS} \cdot Q_{tot}$$

The average Δp_{LS} is around 2 MPa (290 psi). If the pump flow is high the extra loss can be considerable. The power loss also increases if the load pressures vary a lot. The cylinder areas, motor displacements and mechanical torque arms must be designed to match load pressure in order to bring down the power losses. Pump pressure always equals the maximum load pressure when several functions are run simultaneously and the power input to the pump equals the (max. load pressure + Δp_{LS}) x sum of flow.

Five Basic Types of Load-Sensing Systems

- Load sensing *without compensators* in the directional valves. Hydraulically controlled LS-pump.
- Load sensing *with up-stream compensator* for each connected directional valve. Hydraulically controlled LS-pump.
- Load sensing *with down-stream compensator* for each connected directional valve. Hydraulically controlled LS-pump.
- Load sensing *with a combination of up-stream and down-stream compensators*. Hydraulically controlled LS-pump.
- Load sensing with synchronized, both electric controlled pump displacement and electric controlled valve flow area for faster response, increased stability and fewer system losses. This is a new type of LS-system, not yet fully developed.

Technically the down-stream mounted compensator in a valveblock can physically be mounted "up-stream", but work as a down-stream compensator.

System type (3) gives the advantage that activated functions are synchronized independent of pump flow capacity. The flow relation between 2 or more activated functions remains independent of load pressures, even if the pump reaches the maximum swivel angle. This feature is important for machines that often run with the pump at maximum swivel angle and with several activated functions that must be synchronized in speed, such as with excavators. With type (4) system, the functions with *up-stream* compensators have priority. Example: Steering-function for a wheel loader. The system type with down-stream compensators usually have a unique trademark depending on the manufacturer of the valves, for example "LSC" (Linde Hydraulics), "LUDV" (Bosch Rexroth Hydraulics) and "Flowsharing" (Parker Hydraulics) etc. No official standardized name for this type of system has been established but Flowsharing is a common name for it.

Open and Closed Circuits

Open-loop: Pump-inlet and motor-return (via the directional valve) are connected to the hydraulic tank. The term loop applies to feedback; the more correct term is open versus closed "circuit". Open centre circuits use pumps which supply a continuous flow. The flow is returned to tank through the control valve's open centre; that is, when the control valve is centred, it provides an open return path to tank and the fluid is not pumped to a high pressure. Otherwise, if the control valve is actuated it routes fluid to and from an actuator and tank. The fluid's pressure will rise to meet any resistance, since the pump has a constant output. If the pressure rises too high, fluid returns to tank through a pressure relief valve. Multiple control valves may be stacked in series. This type of circuit can use inexpensive, constant displacement pumps.

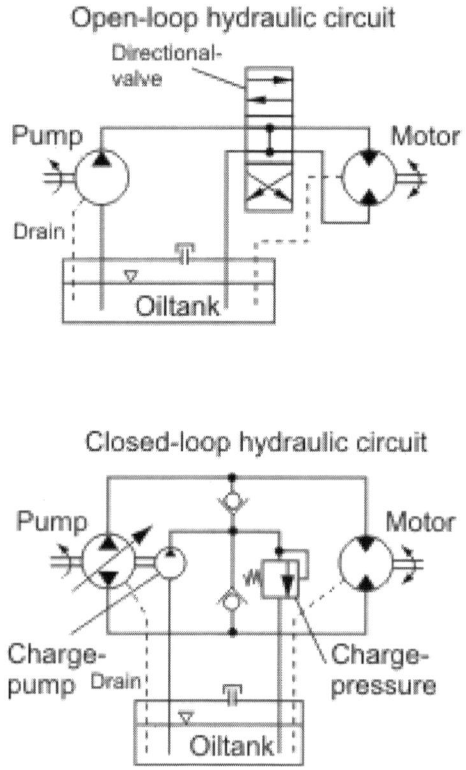

Figure: *Open loop and closed loop circuits*

Closed-loop: Motor-return is connected directly to the pump-inlet. To keep up pressure on the low pressure side, the circuits have a charge pump (a small gearpump) that supplies cooled and filtered oil to the low pressure side. Closed-loop circuits are generally used for hydrostatic transmissions in mobile applications. *Advantages:* No directional valve and better response, the circuit can work with higher pressure. The pump swivel angle covers both positive and negative flow direction. *Disadvantages:* The pump cannot be utilised for any other hydraulic function in an easy way and cooling can be a problem due to limited exchange of oil flow. High power closed loop systems generally must have a 'flush-valve' assembled in the circuit in order to exchange much more flow than the basic leakage flow from the pump and the motor, for increased cooling and filtering. The flush valve is normally integrated in the motor housing to get a cooling effect for the oil that is rotating in the motor housing itself. The losses in the motor housing from rotating effects and losses in the ball bearings can be considerable as motor speeds will reach 4000-5000 rev/min or even more at maximum vehicle speed.

The leakage flow as well as the extra flush flow must be supplied by the charge pump. A large charge pump is thus very important if the transmission is designed for high pressures and high motor speeds. High oil temperature is usually a major problem when using hydrostatic transmissions at high vehicle speeds for longer periods, for instance when transporting the machine from one work place to the other. High oil temperatures for long periods will drastically reduce the lifetime of the transmission. To keep down the oil temperature, the system pressure during transport must be lowered, meaning that the minimum displacement for the motor must be limited to a reasonable value. Circuit pressure during transport around 200-250 bar is recommended.

Closed loop systems in mobile equipment are generally used for the transmission as an alternative to mechanical and hydrodynamic (converter) transmissions. The advantage is a stepless gear ratio (continuously variable speed/torque) and a more flexible control of the gear ratio depending on the load and operating conditions. The hydrostatic transmission is generally limited to around 200 kW maximum power, as the total cost gets too high at higher power compared to a hydrodynamic transmission. Large wheel loaders for instance and heavy machines are therefore usually equipped with converter transmissions. Recent technical achievements for the converter transmissions have improved the efficiency and developments in the software have also improved the characteristics, for example selectable gear shifting programs during operation and more gear steps, giving them characteristics close to the hydrostatic transmission.

Hydrostatic transmissions for earth moving machines, such as for track loaders, are often equipped with a separate 'inch pedal' that is used to temporarily increase the diesel engine rpm while reducing the vehicle speed in order to increase the available hydraulic power output for the working hydraulics at low speeds and increase the tractive effort. The function is similar to stalling a converter gearbox at high engine rpm. The inch function affects the preset characteristics for the 'hydrostatic' gear ratio versus diesel engine rpm.

Components: Hydraulic Pump

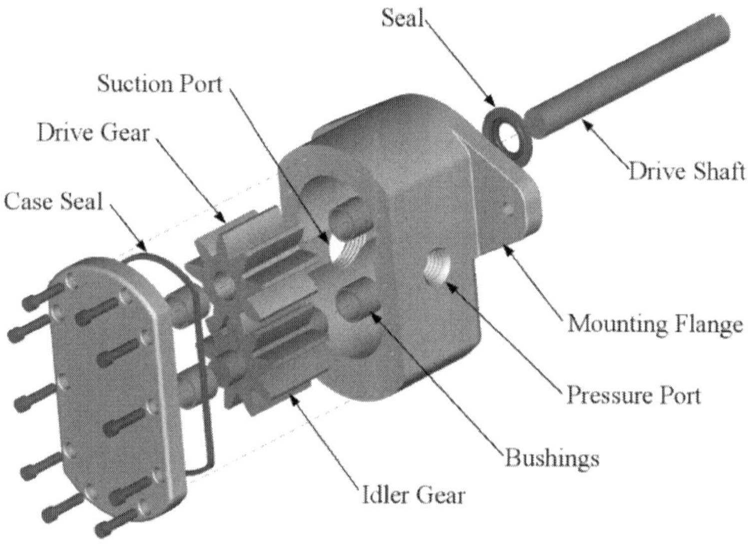

Figure: *An exploded view of an external gear pump.*

Hydraulic pumps supply fluid to the components in the system. Pressure in the system develops in reaction to the load. Hence, a pump rated for 5,000 psi is capable of maintaining flow against a load of 5,000 psi.

Pumps have a power density about ten times greater than an electric motor (by volume). They are powered by an electric motor or an engine, connected through gears, belts, or a flexible elastomeric coupling to reduce vibration.

Common types of hydraulic pumps to hydraulic machinery applications are;

- *Gear pump:* cheap, durable (especially in g-rotor form), simple. Less efficient, because they are constant (fixed) displacement, and mainly suitable for pressures below 20 MPa (3000 psi).
- *Vane pump:* cheap and simple, reliable. Good for higher-flow low-pressure output.

- *Axial piston pump:* many designed with a variable displacement mechanism, to vary output flow for automatic control of pressure. There are various axial piston pump designs, including swashplate (sometimes referred to as a valveplate pump) and checkball (sometimes referred to as a wobble plate pump). The most common is the swashplate pump. A variable-angle swashplate causes the pistons to reciprocate a greater or lesser distance per rotation, allowing output flow rate and pressure to be varied (greater displacement angle causes higher flow rate, lower pressure, and vice versa).
- *Radial piston pump:* normally used for very high pressure at small flows.

Piston pumps are more expensive than gear or vane pumps, but provide longer life operating at higher pressure, with difficult fluids and longer continuous duty cycles. Piston pumps make up one half of a hydrostatic transmission.

Control Valves

Directional control valves route the fluid to the desired actuator. They usually consist of a spool inside a cast iron or steel housing. The spool slides to different positions in the housing, and intersecting grooves and channels route the fluid based on the spool's position.

The spool has a central (neutral) position maintained with springs; in this position the supply fluid is blocked, or returned to tank. Sliding the spool to one side routes the hydraulic fluid to an actuator and provides a return path from the actuator to tank. When the spool is moved to the opposite direction the supply and return paths are switched. When the spool is allowed to return to neutral (centre) position the actuator fluid paths are blocked, locking it in position.

Directional control valves are usually designed to be stackable, with one valve for each hydraulic cylinder, and one fluid input supplying all the valves in the stack.

Tolerances are very tight in order to handle the high pressure and avoid leaking, spools typically have a clearance with the housing of less than a thousandth of an inch (25 μm). The valve block will be mounted to the machine's frame with a *three point* pattern to avoid distorting the valve block and jamming the valve's sensitive components.

The spool position may be actuated by mechanical levers, hydraulic *pilot* pressure, or solenoids which push the spool left or right. A seal allows part of the spool to protrude outside the housing, where it is accessible to the actuator.

The main valve block is usually a stack of *off the shelf* directional control valves chosen by flow capacity and performance. Some valves are designed to be proportional (flow rate proportional to valve position), while others may be simply on-off.

The control valve is one of the most expensive and sensitive parts of a hydraulic circuit.

- Pressure relief valves are used in several places in hydraulic machinery; on the return circuit to maintain a small amount of pressure for brakes, pilot lines, etc... On hydraulic cylinders, to prevent overloading and hydraulic line/seal rupture. On the hydraulic reservoir, to maintain a small positive pressure which excludes moisture and contamination.
- Pressure regulators reduce the supply pressure of hydraulic fluids as needed for various circuits.
- Sequence valves control the sequence of hydraulic circuits; to ensure that one hydraulic cylinder is fully extended before another starts its stroke, for example.
- Shuttle valves provide a logical or function.
- Check valves are one-way valves, allowing an accumulator to charge and maintain its pressure after the machine is turned off, for example.
- Pilot controlled Check valves are one-way valve that can be opened (for both directions) by a foreign pressure signal. For instance if the load should not be held by the check valve anymore. Often the foreign pressure comes from the other pipe that is connected to the motor or cylinder.
- Counterbalance valves are in fact a special type of pilot controlled check valve. Whereas the check valve is open or closed, the counterbalance valve acts a bit like a pilot controlled flow control.
- Cartridge valves are in fact the inner part of a check valve; they are *off the shelf* components with a standardized envelope, making them easy to populate a proprietary valve block. They are available in many configurations; on/off, proportional, pressure relief, etc. They generally screw into a valve block and are electrically controlled to provide logic and automated functions.

- Hydraulic fuses are in-line safety devices designed to automatically seal off a hydraulic line if pressure becomes too low, or safely vent fluid if pressure becomes too high.
- Auxiliary valves in complex hydraulic systems may have auxiliary valve blocks to handle various duties unseen to the operator, such as accumulator charging, cooling fan operation, air conditioning power, etc. They are usually custom valves designed for the particular machine, and may consist of a metal block with ports and channels drilled. Cartridge valves are threaded into the ports and may be electrically controlled by switches or a microprocessor to route fluid power as needed.

Actuators

- Hydraulic cylinder
- Swashplates are used in 'hydraulic motors' requiring highly accurate control and also in 'no stop' continuous (360°) precision positioning mechanisms. These are frequently driven by several hydraulic pistons acting in sequence.
- Hydraulic motor (a pump plumbed in reverse)
- Hydrostatic transmission
- Brakes

Reservoir

The hydraulic fluid reservoir holds excess hydraulic fluid to accommodate volume changes from: cylinder extension and contraction, temperature driven expansion and contraction, and leaks. The reservoir is also designed to aid in separation of air from thc fluid and also work as a heat accumulator to cover losses in the system when peak power is used. Design engineers are always pressured to reduce the size of hydraulic reservoirs, while equipment operators always appreciate larger reservoirs. Reservoirs can also help separate dirt and other particulate from the oil, as the particulate will generally settle to the bottom of the tank. Some designs include dynamic flow channels on the fluid's return path that allow for a smaller reservoir.

Accumulators

Accumulators are a common part of hydraulic machinery. Their function is to store energy by using pressurized gas. One type is a tube with a floating piston. On one side of the piston is a charge of pressurized gas, and on the other side is the fluid. Bladders are used in other designs. Reservoirs store a system's fluid.

Examples of accumulator uses are backup power for steering or brakes, or to act as a shock absorber for the hydraulic circuit.

Hydraulic Fluid

Also known as *tractor fluid*, hydraulic fluid is the life of the hydraulic circuit. It is usually petroleum oil with various additives. Some hydraulic machines require fire resistant fluids, depending on their applications. In some factories where food is prepared, either an edible oil or water is used as a working fluid for health and safety reasons. In addition to transferring energy, hydraulic fluid needs to lubricate components, suspend contaminants and metal filings for transport to the filter, and to function well to several hundred degrees Fahrenheit or Celsius.

Filters

Filters are an important part of hydraulic systems. Metal particles are continually produced by mechanical components and need to be removed along with other contaminants.

Filters may be positioned in many locations. The filter may be located between the reservoir and the pump intake. Blockage of the filter will cause cavitation and possibly failure of the pump. Sometimes the filter is located between the pump and the control valves. This arrangement is more expensive, since the filter housing is pressurized, but eliminates cavitation problems and protects the control valve from pump failures. The third common filter location is just before the return line enters the reservoir. This location is relatively insensitive to blockage and does not require a pressurized housing, but contaminants that enter the reservoir from external sources are not filtered until passing through the system at least once.

Tubes, Pipes and Hoses

Hydraulic tubes are seamless steel precision pipes, specially manufactured for hydraulics. The tubes have standard sizes for different pressure ranges, with standard diameters up to 100 mm. The tubes are supplied by manufacturers in lengths of 6 m, cleaned, oiled and plugged. The tubes are interconnected by different types of flanges (especially for the larger sizes and pressures), welding cones/nipples (with o-ring seal), several types of flare connection and by cut-rings. In larger sizes, hydraulic pipes are used. Direct joining of tubes by welding is not acceptable since the interior cannot be inspected.

Hydraulic pipe is used in case standard hydraulic tubes are not available. Generally these are used for low pressure. They can be connected by threaded connections, but usually by welds. Because of the larger diameters the pipe can usually be inspected internally after welding. Black pipe is non-galvanized and suitable for welding.

Hydraulic hose is graded by pressure, temperature, and fluid compatibility. Hoses are used when pipes or tubes can not be used, usually to provide flexibility for machine operation or maintenance. The hose is built up with rubber and steel layers. A rubber interior is surrounded by multiple layers of woven wire and rubber. The exterior is designed for abrasion resistance. The bend radius of hydraulic hose is carefully designed into the machine, since hose failures can be deadly, and violating the hose's minimum bend radius will cause failure. Hydraulic hoses generally have steel fittings swaged on the ends. The weakest part of the high pressure hose is the connection of the hose to the fitting. Another disadvantage of hoses is the shorter life of rubber which requires periodic replacement, usually at five to seven year intervals. Tubes and pipes for hydraulic applications are internally oiled before the system is commissioned. Usually steel piping is painted outside. Where flare and other couplings are used, the paint is removed under the nut, and is a location where corrosion can begin. For this reason, in marine applications most piping is stainless steel.

Seals, Fittings and Connections

In general, valves, cylinders and pumps have female threaded bosses for the fluid connection, and hoses have female ends with captive nuts. A male-male fitting is chosen to connect the two. Many standardized systems are in use.

Fittings serve several purposes;

1. To bridge different standards; O-ring boss to JIC, or pipe threads to face seal, for example.
2. To allow proper orientation of components, a 90°, 45°, straight, or swivel fitting is chosen as needed. They are designed to be positioned in the correct orientation and then tightened.
3. To incorporate bulkhead hardware.
4. A *quick disconnect* fitting may be added to a machine without modification of hoses or valves

A typical piece of heavy equipment may have thousands of sealed connection points and several different types:

- Pipe fittings, the fitting is screwed in until tight, difficult to orient an angled fitting correctly without over or under tightening.

- O-ring boss, the fitting is screwed into a boss and orientated as needed, an additional nut tightens the fitting, washer and o-ring in place.
- Flare fittings, are metal to metal compression seals deformed with a cone nut and pressed into a flare mating.
- Face seal, metal flanges with a groove and o-ring are fastened together.
- Beam seals are costly metal to metal seals used primarily in aircraft.
- Swaged seals, tubes are connected with fittings that are swaged permanently in place. Primarily used in aircraft.

Elastomeric seals (O-ring boss and face seal) are the most common types of seals in heavy equipment and are capable of reliably sealing 6000+ psi (40+ MPa) of fluid pressure.

Basic Calculations

Hydraulic power is defined as flow times pressure. The hydraulic power supplied by a pump:

$$\text{Power} = (P \times Q) \div 600$$

where power is in kilowatts [kW], P pressure in bars, and Q is the flow in liters per minute. For example, a pump delivers 180 lit/min and the pressure equals 250 bar, therefore the power of the pump is 75 kW.

When calculating the power input to the pump, the total pump efficiency η_{total} must be included. This efficiency is the product of volumetric efficiency, η_{vol} and the hydromechanical efficiency, η_{hm}. Power input = Power output $\div$ η_{total}. The average for axial piston pumps, $\eta_{total} = 0.87$. In the example the power source, for example a diesel engine or an electric motor, must be capable of delivering at least $75 \div 0.87 = 86$ [kW]. The hydraulic motors and cylinders that the pump supplies with hydraulic power also have efficiencies and the total system efficiency (without including the pressure drop in the hydraulic pipes and valves) will end up at approx. 0.75. Cylinders normally have a total efficiency around 0.95 while hydraulic axial piston motors 0.87, the same as the pump. In general the power loss in a hydraulic energy transmission is thus around 25% or more at ideal viscosity range 25-35 [cSt].

Calculation of the required max. power output for the diesel engine, rough estimation:

(1) Check the max. powerpoint, i.e. the point where pressure times flow reach the max. value.

(2) $E_{diesel} = (P_{max} \cdot Q_{tot}) \div \eta$.

Q_{tot} = calculate with the theoretical pump flow for the consumers not including leakages at max. power point.

P_{max} = actual pump pressure at max. power point.

Note: η is the total efficiency = (output mechanical power ÷ input mechanical power). For rough estimations, $\eta = 0.75$. Add 10-20% (depends on the application) to this power value.

(3) Calculate the required pumpdisplacement from required max. sum of flow for the consumers in worst case and the diesel engine rpm in this point. The max. flow can differ from the flow used for calculation of the diesel engine power. Pump volumetric efficiency average, piston pumps: $\eta_{vol} = 0.93$.

Pumpdisplacement $V_{pump} = Q_{tot} \div n_{diesel} \chi\ 0.93$.

(4) Calculation of prel. cooler capacity: Heat dissipation from hydraulic oil tanks, valves, pipes and hydraulic components is less than a few percent in standard mobile equipment and the cooler capacity must include some margins. Minimum cooler capacity, $E_{cooler} = 0.25E_{diesel}$

At least 25% of the input power must be dissipated by the cooler when peak power is utilised for long periods. In normal case however, the peak power is used for only short periods, thus the actual cooler capacity required might be considerably less. The oil volume in the hydraulic tank is also acting as a heat accumulator when peak power is used. The system efficiency is very much dependent on the type of hydraulic work tool equipment, the hydraulic pumps and motors used and power input to the hydraulics may vary a lot.

Each circuit must be evaluated and the load cycle estimated. New or modified systems must always be tested in practical work, covering all possible load cycles. An easy way of measuring the actual average power loss in the system is to equip the machine with a test cooler and measure the oil temperature at cooler inlet, oil temperature at cooler outlet and the oil flow through the cooler, when the machine is in normal operating mode.

From these figures the test cooler power dissipation can be calculated and this is equal to the power loss when temperatures are stabilized. From this test the actual required cooler can be calculated to reach specified oil temperature in the oil tank. One problem can be to assemble the measuring equipment inline, especially the oil flow metre.

History

Joseph Bramah patented the hydraulic press in 1795. While working at Bramah's shop, Henry Maudslay suggested a cup leather packing. Because it produced superior results, the hydraulic press eventually displaced the steam hammer from metal forging.

To supply small scale power that was impractical for individual steam engines, central station hydraulic systems were developed. Hydraulic power was used to operate cranes and other machinery in British ports and elsewhere in Europe. The largest hydraulic system was in London. Hydraulic power was used extensively in Bessemer steel production. Hydraulic power was also used for elevators, to operate canal locks and rotating sections of bridges. Some of these systems remained in use well into the twentieth century.

Harry Franklin Vickers was called the "Father of Industrial Hydraulics" by ASME.

Automatic Transmission

An automatic transmission (also called automatic gearbox) is a type of motor vehicle transmission that can automatically change gear ratios as the vehicle moves, freeing the driver from having to shift gears manually. Most automatic transmissions have a defined set of gear ranges, often with a parking pawl feature that locks the output shaft of the transmission stroke face.

Similar but larger devices are also used for heavy-duty commercial and industrial vehicles and equipment. Some machines with limited speed ranges or fixed engine speeds, such as some forklifts and lawn mowers, only use a torque converter to provide a variable gearing of the engine to the wheels.

Besides automatics, there are also other types of automated transmissions such as a continuously variable transmission (CVT) and semi-automatic transmissions, that free the driver from having to shift gears manually, by using the transmission's computer to change gear, if for example the driver were redlining the engine. Despite superficial similarity to other transmissions, automatic transmissions differ significantly in internal operation and driver's feel from semi-automatics and CVTs. An automatic uses a torque converter instead of a clutch to manage the connection between the transmission gearing and the engine. In contrast, a CVT uses a belt or other torque transmission scheme to allow an "infinite" number of gear ratios instead of a fixed number of gear ratios. A semi-automatic retains a clutch like a manual transmission, but controls the clutch through electrohydraulic means.

A conventional manual transmission is frequently the base equipment in a car, with the option being an automated transmission such as a conventional automatic, semi-automatic, or CVT. The ability to shift gears manually, often via paddle shifters, can also be found on certain automated transmissions (manumatics such as Tiptronic), semi-automatics (BMW SMG), and CVTs (such as Lineartronic).

The first automatic transmission was invented in 1921 by Alfred Horner Munro of Regina, Saskatchewan, Canada, and patented under Canadian patent CA 235757 in 1923. (Munro obtained UK patent GB215669 215,669 for his invention in 1924 and US patent 1,613,525 on 4 January 1927). Being a steam engineer, Munro designed his device to use compressed air rather than hydraulic fluid, and so it lacked power and never found commercial application. The first automatic transmissions using hydraulic fluid were developed by General Motors during the 1930s and introduced in the 1940 Oldsmobile as the "Hydra-Matic" transmission. They were incorporated into GM-built tanks during World War II and, after the war, GM marketed them as being "battle-tested".

Comparison with Manual Transmission

Most cars sold in North America since the 1950s have been available with an automatic transmission. Conversely, in Europe a manual gearbox is standard, with 20% of drivers opting for an automatic transmission. In some Asian markets and in Australia, automatic transmissions have become very popular since the 1990s.

Vehicles equipped with automatic transmissions are not as complex to drive. Consequently, in some jurisdictions, drivers who have passed their driving test in a vehicle with an automatic transmission will not be licensed to drive a manual transmission vehicle. Conversely, a manual license will allow the driver to drive both manual and automatic vehicles. Examples of driving license restrictions are Croatia, Dominican Republic, Israel, United Kingdom, Brazil, some states in Australia, France, Portugal, Latvia, Lebanon, Lithuania, Ireland, Belgium, Germany, the Netherlands, Sweden, Spain, Austria, Norway, Hungary, South Africa, Trinidad and Tobago, Belize, Japan, China,

Hong Kong, Macau, Mauritius, South Korea, Romania, Singapore, Philippines, United Arab Emirates, India, Estonia, Finland, Saudi Arabia (in March 2012), Switzerland, Slovenia, Republic of Ireland and New Zealand (restricted licence only).

Automatic Transmission Modes

Conventionally, in order to select the transmission operating mode, the driver moves a selection lever located either on the steering column or on the floor (as with a manual on the floor, except that most automatic selectors on the floor do not move in the same type of pattern as a manual lever; most automatic levers only move vertically). In order to select modes, or to manually select specific gear ratios, the driver must push a button in (called the shift lock button) or pull the handle (only on column mounted shifters) out.

Some vehicles position selector buttons for each mode on the cockpit instead, freeing up space on the central console. Vehicles conforming to US Government standards must have the modes ordered P-R-N-D-L (left to right, top to bottom, or clockwise). Prior to this, quadrant-selected automatic transmissions often used a P-N-D-L-R layout, or similar. Such a pattern led to a number of deaths and injuries owing to driver error causing unintentional gear selection, as well as the danger of having a selector (when worn) jump into Reverse from Low gear during engine braking maneuvers.

Automatic transmissions have various modes depending on the model and make of the transmission. Some of the common modes include:

Park (P)

This selection mechanically locks the output shaft of transmission, restricting the vehicle from moving in any direction. A parking pawl prevents the transmission from rotating, and therefore the vehicle from moving, but the vehicle's driven wheels may still rotate individually (because of the differential), as well as the non-driven roadwheels may still rotate freely. For this reason, it is recommended to use the hand brake (*parking brake*) because this actually locks (in most cases) the wheels and prevents them from moving. This also increases the life of the transmission and the park pin mechanism, because parking on an incline with the transmission in park without the parking brake engaged will cause undue stress on the parking pin. A hand brake should also prevent the car from moving if a worn selector accidentally drops into reverse gear while idleing.

It should be noted that locking the transmission output shaft with park does not definitivly lock the driving wheels. If one driving wheel has little vertical load it will tend to slip, and will rotate in the opposite direction to the more heavily loaded non-slipping wheel. Only a parking brake can be relied upon to positively lock both of the parking-braked wheels. (This is not the case with certain 1950's Chrysler products that carried their parking brake on the transmission tailshaft, a defect compounded by the provision of a bumper jack.) It is typical of front-wheel-drive vehicles for the parking brake to lock the rear (non-driving) wheels, so use of both the parking brake and the transmission park lock provides the greatest security against unintended movement on slopes. Buses in Hong Kong *(Kowloon Motor Bus)* do not have Parking mode (P), instead, they are using a single lever to prevent the whole bus from moving.

A car should be allowed to come to a complete stop before setting the transmission into park to prevent damage. Usually, Park (P) is one of only two selections in which the car's engine can be started, the other being Neutral (N) (Buses in Hong Kong mentioned above has to be in Neutral mode (N) ONLY before the engine can be started). In many modern cars and trucks, the driver must have the foot brake applied before the transmission can be taken out of park. The Park position is omitted on buses/coaches (and some road tractors) with automatic transmission (on which a parking pawl is not practical), which must be placed in neutral with the air-operated parking brakes set. Advice is given in some owner's manuals that if the vehicle is parked on a steep slope using the park lock only, it may not be possible to release the park lock (move the selector lever out of P). Another vehicle may be required to push the stuck vehicle uphill slightly to remove the loading on the park lock pawl.

Most automobiles require P or N to be set on the selector lever before the engine can be started. This is typically achieved via a normally-open inhibitor switch (sometimes called a "neutral safety switch") wired in series with the starter motor engagement circuit, which is closed when P or N is selected, completing the circuit (when the key is turned to the start position), along with any other safety devices which may be present on newer cars (such as a foot-brake application).

Reverse (R)

This engages reverse gear within the transmission, permitting the vehicle to be driven backward, and operates a switch to turn on the backup lights for improved visibility (the switch may also activate a beeper on delivery trucks or other large vehicles to warn other drivers and nearby pedestrians of the driver's reverse movement). To select reverse in most transmissions, the driver must come to a complete stop, depress the shift lock button (or move the shift lever toward the driver in a column shifter, or move the shifter sideways along a notched channel in a console shifter) and select reverse. Not coming to a complete stop may cause severe damage to the transmission.

Some modern automatic transmissions have a safety mechanism in place, which does, to some extent, prevent (but not completely avoid) inadvertently putting the car in reverse when the vehicle is moving forward; such a mechanism may consist of a solenoid-controlled physical barrier on either side of the Reverse position, electronically engaged by a switch on the brake pedal. Therefore, the brake pedal needs to be depressed in order to allow the selection of reverse. Some electronic transmissions prevent or delay engagement of reverse gear altogether while the car is moving.

Some shifters with a shift button allow the driver to freely move the shifter from R to N or D, or simply moving the shifter to N or D without actually depressing the button. However, the driver cannot shift back to R without depressing the shift button, to prevent accidental shifting, especially at high speeds, which could damage the transmission.

Neutral / No Gear (N)

This disengages all gear trains within the transmission, effectively disconnecting the transmission from the driven wheels, allowing the vehicle to coast freely under its own weight and gain momentum without the motive force from the engine. Coasting in idle down long grades (where law permits) should be avoided, though, as the transmission's lubrication pump is driven by non-idle engine RPMs. Similarly, emergency towing with an automatic transmission in neutral should be a last resort. Manufacturers understand emergency situations and list limitations of towing a vehicle in neutral (usually not to exceed 55 mph and 50 miles). This is the only other selection in which the vehicle's engine may be started.

Drive (D)

This position allows the transmission to engage the full range of available forward gear ratios, allowing the vehicle to move forward and accelerate through its range of gears. The number of gear ratios within the transmission depends on the model, but they initially ranged from three (predominant before the 1990s), to four and five speeds (losing popularity to six-speed autos, though still favoured by Chrysler and Honda/Acura). Six-speed automatic transmissions are probably the most common offering in cars and trucks from 2010 in carmakers as Toyota, GM and Ford. However, seven-speed automatics are becoming available in some high-performance production luxury cars (found in Mercedes 7G gearbox, Infiniti), as are eight-speed autos in models from 2006 introduced by Aisin Seiki Co. in Lexus, ZF and Hyundai Motor Company. From 2013 are available nine speeds transmissions produced by ZF and Mercedes 9G.

Overdrive ('D', 'OD', or a boxed [D] or the absence of an illuminated 'O/D OFF')

This mode is used in some transmissions to allow early computer-controlled transmissions to engage the automatic overdrive. In these transmissions, Drive (D) locks the automatic overdrive off, but is identical otherwise. OD (Overdrive) in these cars is engaged under steady speeds or low acceleration at approximately 35–45 mph (56–72 km/h). Under hard acceleration or below 35–45 mph (56–72 km/h), the transmission will automatically downshift. Other vehicles with this selector (example light trucks) will not only disable up-shift to the overdrive gear, but keep the remaining available gears continuously engaged to the engine for use of compression braking. Verify the behaviour of this switch and consider the benefits of reduced friction brake use when city driving where speeds typically do not necessitate the overdrive gear.

Third (3)

This mode limits the transmission to the first three gear ratios, or sometimes locks the transmission in third gear. This can be used to climb or going down hill. Some vehicles will automatically shift up out of third gear in this mode if a certain revolutions per minute (RPM) range is reached in order to prevent engine damage. This gear is also recommended while towing a trailer.

Second (2 or S)

This mode limits the transmission to the first two gear ratios, or locks the transmission in second gear on Ford, Kia, and Honda models. This can be used to drive in adverse conditions such as snow and ice, as well as climbing or going down hills in winter. It is usually recommended to use second gear for starting on snow and ice, and use of this position enables this with an automatic transmission. Some vehicles will automatically shift up out of second gear in this mode if a certain RPM range is reached in order to prevent engine damage.

Although traditionally considered second gear, there are other names used. Chrysler models with a three-speed automatic since the late 1980s have called this gear 3 while using the traditional names for *Drive* and *Low*. Oldsmobile has called second gear as the 'Super' range — which was first used on their 4-speed Hydramatic transmissions, although the use of this term continued until the early 1980s when GM's Turbo Hydramatic automatic transmissions were standardized by all of their divisions years after the 4-speed Hydramatic was discontinued.

First (1 or L [Low])

This mode locks the transmission in first gear only. In older vehicles, it will not change to any other gear range. Some vehicles will automatically shift up out of first gear in this mode if a certain RPM range is reached in order to prevent engine damage. This, like second, can be used during the winter season, for towing, or for downhill driving to increase the engine braking effect. As well as the above modes there are also other modes, dependent on the manufacturer and model. Some examples include:

D5

In Hondas and Acuras equipped with five-speed automatic transmissions, this mode is used commonly for highway use (as stated in the manual), and uses all five forward gears.

D4

This mode is also found in Honda and Acura four or five-speed automatics, and only uses the first four gear ratios. According to the manual, it is used for stop-and-go traffic, such as city driving.

D3 or 3

This mode is found in Honda, Acura, Volkswagen and Pontiac four-speed automatics and only uses the first three gear ratios. According to the manual, it is used for stop-and-go traffic, such as city driving.

D2 and D1

These modes are found on older Ford transmissions (C6, etc.). In D1, all three gears are used, whereas in D2 the car starts in second gear and upshifts to third.

S or Sport

This is commonly described as Sport mode. It operates in an identical manner as "D" mode, except that the upshifts change much higher up the engine's rev range. This has the effect on maximising all the available engine output, and therefore enhances the performance of the vehicle, particularly during acceleration. This mode will also downchange much higher up the rev range compared to "D" mode, maximising the effects of engine braking. This mode will have a detrimental effect on fuel economy. Hyundai has a Norm/Power switch next to the gearshift for this purpose on the *Tiburon.*

Some early GMs equipped with HYDRA-MATIC transmissions used (S) to indicate Second gear, being the same as the 2 position on a Chrysler, shifting between only first and second gears. This would have been recommended for use on steep grades, or slippery roads like dirt, or ice, and limited to speeds under 40 mph. (L) was used in some early GMs to indicate (L)ow gear, being the same as the 2 position on a Chrysler, locking the transmission into first gear. This would have been recommended for use on steep grades, or slippery roads like dirt, or ice, and limited to speeds under 15 mph.

+ –, and M

This is for the Manual mode selection of gears in certain automatics, such as Porsche's Tiptronic and Honda's StepTronic. The M feature can also be found in Chrysler and General Motors products such as the Dodge Magnum, Journey, and Pontiac G6, Mazda products such as the Mazda 3, Mazda6, and the CX-7, as well as Toyota's Camry, Corolla, Fortuner, Previa and Innova. Mitsubishi and some Audi models (Audi TT), meanwhile do not have the M, and instead have the + and -, which is separated from the rest of the shift modes; the same is true for some Peugeot products like Peugeot 206. Meanwhile, the driver can shift up and down at will by toggling the (console mounted) shift lever similar to a semi-automatic transmission. This mode may be engaged either through a selector/position or by actually changing the gears (e.g., tipping the gear-down paddles mounted near the driver's fingers on the steering wheel).

Winter (W)

In some Volvo, Mercedes-Benz, BMW and General Motors Europe models, a winter mode can be engaged so that second gear is selected instead of first when pulling away from stationary, to reduce the likelihood of loss of traction due to wheel spin on snow or ice. On GM cars, this was D2 in the 1950s, and is Second Gear Start after 1990. On Ford, Kia, and Honda automatics, this feature can be accessed by moving the gear selector to 2 to start, then taking your foot off the accelerator while selecting D once the car is moving.

Brake (B)

A mode selectable on some Toyota models. In non-hybrid cars, this mode lets the engine do compression braking, also known as engine braking, typically when encountering a steep downhill. Instead of engaging the brakes, the engine in a non-hybrid car switches to a lower gear and slows down the spinning tires. The engine holds the car back, instead of the brakes slowing it down. GM called this "HR" ("hill retarder") and "GR" ("grade retarder") in the 1950s. For hybrid cars, this mode converts the electric motor into a generator for the battery (Regenerative Braking). It is not the same as downshifting in a non-hybrid car, but it has the same effect in slowing the car without using the brakes.

Hydraulic Automatic Transmissions

The predominant form of automatic transmission is hydraulically operated; using a fluid coupling or torque converter, and a set of planetary gearsets to provide a range of gear ratios.

Parts and Operation

A hydraulic automatic transmission consists of the following parts:

- *Torque converter:* A type of fluid coupling, hydraulically connecting the engine to the transmission. It takes the place of a mechanical clutch, allowing the transmission to stay in gear and the engine to remain running while the vehicle is stationary, without stalling. A torque converter differs from a fluid coupling, in that it provides a variable amount of torque multiplication at low engine speeds, increasing breakaway acceleration. This is accomplished with a third member in the coupling assembly known as the stator, and by altering the shapes of the vanes inside the coupling in such a way as to curve the fluid's path into the stator. The stator captures the kinetic energy of the transmission fluid, in effect using the leftover force of it to enhance torque multiplication.
- *Pump*: Not to be confused with the impeller inside the torque converter, the pump is typically a gear pump mounted between the torque converter and the planetary gearset. It draws transmission fluid from a sump and pressurizes it, which is

needed for transmission components to operate. The input for the pump is connected to the torque converter housing, which in turn is bolted to the engine's flywheel, so the pump provides pressure whenever the engine is running and there is enough transmission fluid. Early automatic transmissions also had a rear pump, allowing push-starting.

- *Planetary gearset*: A compound epicyclic planetary gearset, whose bands and clutches are actuated by hydraulic servos controlled by the valve body, providing two or more gear ratios. (Not part of some manufacturers transmissions during some eras, Honda being but one).
- *Clutches and bands:* to effect gear changes, one of two types of clutches or bands are used to hold a particular member of the planetary gearset motionless, while allowing another member to rotate, thereby transmitting torque and producing gear reductions or overdrive ratios. These clutches are actuated by the valve body, their sequence controlled by the transmission's internal programming. Principally, a type of device known as a sprag or roller clutch is used for routine upshifts/downshifts. Operating much as a ratchet, it transmits torque only in one direction, free-wheeling or "overrunning" in the other. The advantage of this type of clutch is that it eliminates the sensitivity of timing a simultaneous clutch release/apply on two planetaries, simply "taking up" the drivetrain load when actuated, and releasing automatically when the next gear's sprag clutch assumes the torque transfer. The bands come into play for manually selected gears, such as low range or reverse, and operate on the planetary drum's circumference. Bands are not applied when drive/overdrive range is selected, the torque being transmitted by the sprag clutches instead. Bands are used for braking; the GM Turbo-Hydramatics incorporated this..
- *Valve body*: hydraulic control centre that receives pressurized fluid from the *main pump* operated by the fluid coupling/ torque converter. The pressure coming from this pump is regulated and used to run a network of spring-loaded valves, check balls and servo pistons. The valves use the pump pressure and the pressure from a centrifugal governor on the output side (as well as hydraulic signals from the range selector valves and the *throttle valve* or *modulator*) to control which ratio is selected on the gearset; as the vehicle and engine

change speed, the difference between the pressures changes, causing different sets of valves to open and close. The hydraulic pressure controlled by these valves drives the various clutch and brake band actuators, thereby controlling the operation of the planetary gearset to select the optimum gear ratio for the current operating conditions. However, in many modern automatic transmissions, the valves are controlled by electro-mechanical servos which are controlled by the electronic engine control unit (ECU) or a separate transmission control unit (TCU, also known as transmission control module (TCM).

- *Hydraulic & lubricating oil*: called automatic transmission fluid (ATF), this component of the transmission provides lubrication, corrosion prevention, and a hydraulic medium to convey mechanical power (for the operation of the transmission). Primarily made from refined petroleum, and processed to provide properties that promote smooth power transmission and increase service life, the ATF is one of the few parts of the automatic transmission that needs routine service as the vehicle ages.

The multitude of parts, along with the complex design of the valve body, originally made hydraulic automatic transmissions much more complicated (and expensive) to build and repair than manual transmissions. In most cars (except US family, luxury, sport-utility vehicle, and minivan models) they have usually been extra-cost options for this reason. Mass manufacturing and decades of improvement have reduced this cost gap.

Energy Efficiency

Hydraulic automatic transmissions are almost always less energy efficient than manual transmissions due mainly to viscous and pumping losses, both in the torque converter and the hydraulic actuators. A relatively small amount of energy is required to pressurize the hydraulic control system, which uses fluid pressure to determine the correct shifting patterns and operate the various automatic clutch mechanisms.

Manual transmissions use a mechanical clutch to transmit torque, rather than a torque converter, thus avoiding the primary source of loss in an automatic transmission. Manual transmissions also avoid the power requirement of the hydraulic control system, by relying on the human muscle power of the vehicle operator to disengage the clutch and actuate the gear levers, and the mental power of the operator to make appropriate gear ratio selections. Thus the manual transmission requires very little engine power to function, with the main power consumption due to drag from the gear train being immersed in the lubricating oil of the gearbox.

The on-road acceleration of an automatic transmission can occasionally exceed that of an otherwise identical vehicle equipped with a manual transmission in turbocharged diesel applications. Turbo-boost is normally lost between gear changes in a manual whereas in an automatic the accelerator pedal can remain fully depressed. This however is still largely dependent upon the number and optimal spacing of gear ratios for each unit, and whether or not the elimination of spooldown/accelerator lift off represent a significant enough gain to counter the slightly higher power consumption of the automatic transmission itself.

History and Improvements

Modern automatic transmissions can trace their origins to an early "horseless carriage" gearbox that was developed in 1904 by the Sturtevant brothers of Boston, Massachusetts. This unit had two forward speeds, the ratio change being brought about by flyweights that were driven by the engine. At higher engine speeds, high gear was engaged. As the vehicle slowed down and engine RPM decreased, the gearbox would shift back to low. Unfortunately, the metallurgy of the time wasn't up to the task, and owing to the abruptness of the gear change, the transmission would often fail without warning.

The next significant phase in the automatic transmission's development occurred in 1908 with the introduction of Henry Ford's remarkable Model T. The Model T, in addition to being cheap and reliable by the standards of the day, featured a simple, two speed plus reverse planetary transmission whose operation was manually controlled by the driver using pedals. The pedals actuated the transmission's friction elements (bands and clutches) to select the desired gear. In some respects, this type of transmission was less demanding of the driver's skills than the contemporary, unsynchronized manual transmission, but still required that the driver know when to make a shift, as well as how to get the car off to a smooth start.

In 1934, both REO and General Motors developed semi-automatic transmissions that were less difficult to operate than a fully manual unit. These designs, however, continued to use a clutch to engage the engine with the transmission. The General Motors unit, dubbed the "Automatic Safety Transmission," was notable in that it employed a power-shifting planetary gearbox that was hydraulically controlled and was sensitive to road speed, anticipating future development. Parallel to the development in the 1930s of an automatically shifting gearbox was Chrysler's work on adapting the fluid coupling to automotive use. Invented early in the 20th century, the fluid coupling was the answer to the question of how to avoid stalling the engine when the vehicle was stopped with the transmission in gear. Chrysler itself never used the fluid coupling with any of its automatic transmissions, but did use it in conjunction with a hybrid manual transmission called "Fluid Drive" (the similar Hy-Drive used a torque converter). These developments in automatic gearbox and fluid coupling technology eventually culminated in the introduction in 1939 of the General Motors Hydra-Matic, the world's first mass-produced automatic transmission.

Available as an option on 1940 Oldsmobiles and later Cadillacs, the Hydra-Matic combined a fluid coupling with three hydraulically-controlled planetary gearsets to produce four forward speeds plus reverse. The transmission was sensitive to engine throttle position and road speed, producing fully automatic up- and down-shifting that varied according to operating conditions.

The Hydra-Matic was subsequently adopted by Cadillac and Pontiac, and was sold to various other automakers, including Bentley, Hudson, Kaiser, Nash, and Rolls-Royce. It also found use during World War II in some military vehicles. From 1950-1954, Lincoln cars were also available with the Hydra-Matic. Mercedes-Benz subsequently devised a four-speed fluid coupling transmission that was similar in principle to the Hydra-Matic, but of a different design.

Interestingly, the original Hydra-Matic incorporated two features which are widely emulated in today's transmissions. The Hydra-Matic's ratio spread through the four gears produced excellent "step-off" and acceleration in first, good spacing of intermediate gears, and the effect of an overdrive in fourth, by virtue of the low numerical rear axle ratio used in the vehicles of the time. In addition, in third and fourth gear, the fluid coupling only handled a portion of the engine's torque, resulting in a high degree of efficiency. In this respect, the transmission's behaviour was similar to modern units incorporating a lock-up torque converter.

In 1956, GM introduced the "Jetaway" Hydra-Matic, which was different in design than the older model. Addressing the issue of shift quality, which was an ongoing problem with the original Hydra-Matic, the new transmission utilised two fluid couplings, the primary one that linked the transmission to the engine, and a secondary one that replaced the clutch assembly that controlled the forward gearset in the original. The result was much smoother shifting, especially from first to second gear, but with a loss in efficiency and an increase in complexity. Another innovation for this new style Hydra-Matic was the appearance of a Park position on the selector. The original Hydra-Matic, which continued in production until the mid-1960s, still used the Reverse position for parking pawl engagement.

The first torque converter automatic, Buick's Dynaflow, was introduced for the 1948 model year. It was followed by Packard's Ultramatic in mid-1949 and Chevrolet's Powerglide for the 1950 model year. Each of these transmissions had only two forward speeds, relying on the converter for additional torque multiplication. In the early 1950s, BorgWarner developed a series of three-speed torque converter automatics for American Motors, Ford Motor Company, Studebaker, and several other manufacturers in the US and other countries. Chrysler was late in developing its own true automatic, introducing the two-speed torque converter PowerFlite in 1953, and the three-speed TorqueFlite in 1956. The latter was the first to utilise the Simpson compound planetary gearset.

General Motors produced multiple-turbine torque converters from 1954 to 1961. These included the Twin-Turbine Dynaflow and the triple-turbine Turboglide transmissions. The shifting took place in the torque converter, rather than through pressure valves and changes in planetary gear connections. Each turbine was connected to the drive shaft through a different gear train. These phased from one ratio to another according to demand, rather than shifting. The Turboglide actually had two speed ratios in reverse, with one of the turbines rotating backwards.

By the late 1960s, most of the fluid-coupling four-speed and two-speed transmissions had disappeared in favour of three-speed units with torque converters. Also around this time, whale oil was removed from automatic transmission fluid. By the early 1980s, these were being supplemented and eventually replaced by overdrive-equipped transmissions providing four or more forward speeds. Many transmissions also adopted the lock-up torque converter (a mechanical clutch locking the torque converter pump and turbine together to eliminate slip at cruising speed) to improve fuel economy.

As computerised engine control units (ECUs) became more capable, much of the logic built into the transmission's valve body was offloaded to the ECU. (Some manufacturers use a separate computer dedicated to the transmission, but sharing information with the engine management computer.) In this case, solenoids turned on and off by the computer control shift patterns and gear ratios, rather than the spring-loaded valves in the valve body. This allows for more precise control of shift points, shift quality, lower shift times, and (on some newer cars) semi-automatic control, where the driver tells the computer when to shift. The result is an impressive combination of efficiency and smoothness. Some computers even identify the driver's style and adapt to best suit it.

ZF Friedrichshafen and BMW were responsible for introducing the first six-speed (the ZF 6HP26 in the 2002 BMW E65 7-Series). Mercedes-Benz's 7G-Tronic was the first seven-speed in 2003, with Toyota introducing an eight-speed in 2007 on the Lexus LS 460. Derived from the 7G-Tronic, Mercedes-Benz unveiled a semi-automatic transmission with the torque converter replaced with a wet multi clutch called the AMG SPEEDSHIFT MCT.

Automatic Transmission Models

Some of the best known automatic transmission families include:

- General Motors — Dynaflow, Powerglide, Turboglide, "Turbo-Hydramatic" TH350, TH400 and 700R4, 4L60-E, 4L80-E, Holden Trimatic
- Ford: Cruise-O-Matic, C4, CD4E, C6, AOD/AODE, E4OD, ATX, AXOD/AX4S/AX4N
- Cummins 68 RFE (fitted to the Ram diesel segment)
- Chrysler: TorqueFlite 727 and 904, A500, A518, 45RFE, 545RFE
- BorgWarner (later Aisin AW)
- ZF Friedrichshafen automatic transmissions
- Mercedes-Benz transmissions
- Allison Transmission
- Voith Voith Turbo
- Aisin AW; Aisin AW is a Japanese automotive parts supplier, known for its automatic transmissions and navigation systems
- Honda
- Nissan/Jatco
- Volkswagen Group — 01M
- Drivetrain Systems International (DSI) — M93, M97 and M74 4-speeds, M78 and M79 6-speeds

- Hyundai Hyundai Powertech — 4F12, 4F16, 4F23 4-Speeds, 5F25, 5F16, 5F23 5-Speeds, 6F17, 6F26, 6F40 6-Speeds, 8R40, 8R50 8-Speeds, Mini Cooper — Automatic or manual transmission, all models

Automatic transmission families are usually based on Ravigneaux, Lepelletier, or Simpson planetary gearsets. Each uses some arrangement of one or two central sun gears, and a ring gear, with differing arrangements of planet gears that surround the sun and mesh with the ring. An exception to this is the Hondamatic line from Honda, which uses sliding gears on parallel axes like a manual transmission without any planetary gearsets. Although the Honda is quite different from all other automatics, it is also quite different from an automated manual transmission (AMT). Many of the above AMTs exist in modified states, which were created by racing enthusiasts and their mechanics by systematically re-engineering the transmission to achieve higher levels of performance. These are known as "performance transmissions". An example of a manufacturer of high performance transmissions of General Motors and Ford transmissions is PerformaBuilt.

Continuously Variable Transmissions

A fundamentally different type of automatic transmission is the *continuously variable transmission* or *CVT*, which can smoothly and steplessly alter its gear ratio by varying the diameter of a pair of belt or chain-linked pulleys, wheels or cones. Some continuously variable transmissions use a hydrostatic drive — consisting of a variable displacement pump and a hydraulic motor — to transmit power without gears. CVT designs are usually as fuel efficient as manual transmissions in city driving, but early designs lose efficiency as engine speed increases.

A slightly different approach to CVT is the concept of *toroidal CVT* or *infinitely variable transmission* (IVT). These concepts provide zero and reverse gear ratios. Some current hybrid vehicles, notably those of Toyota, Lexus and Ford Motor Company, have an electronically controlled CVT (E-CVT). In this system, the transmission has fixed gears, but the ratio of wheel-speed to engine-speed can be continuously varied by controlling the speed of the third input to a differential using an electric motor-generator.

Manually Controlled Automatic Transmissions

Most automatic transmissions offer the driver a certain amount of manual control over the transmission's shifts (beyond the obvious selection of forward, reverse, or neutral). Those controls take several forms:

Throttle Kickdown: Most automatic transmissions include some means of forcing a downshift into the lowest possible gear ratio if the throttle pedal is fully depressed. In many older designs, kickdown is accomplished by mechanically actuating a valve inside the transmission. Most modern designs use a solenoid-operated valve that is triggered by a switch on the throttle linkage or by the engine control unit (ECU) in response to an abrupt increase in engine power.

Mode Selection: Allows the driver to choose between preset shifting programs. For example, Economy mode saves fuel by upshifting at lower engine speeds, while Sport mode (aka "Power" or "Performance") delays upshifting for maximum acceleration. Some transmission units also have Winter mode, where higher gear ratios are chosen to keep revs as low as possible while on slippery surfaces. The modes also change how the computer responds to throttle input.

Low Gear Ranges: Conventionally, automatic transmissions have selector positions that allow the driver to limit the maximum ratio that the transmission may engage. On older transmissions, this was accomplished by a mechanical lockout in the transmission valve body preventing an upshift until the lockout was disengaged; on computer-controlled transmissions, the same effect is accomplished by firmware. The transmission can still upshift and downshift automatically between the remaining ratios: for example, in the *3* range, a transmission could shift from first to second to third, but not into fourth or higher ratios. Some transmissions will still upshift automatically into the higher ratio if the engine reaches its maximum permissible speed in the selected range.

Manual Controls: Some transmissions have a mode in which the driver has full control of ratio changes (either by moving the selector, or through the use of buttons or paddles), completely overriding the automated function of the hydraulic controller. Such control is particularly useful in cornering, to avoid unwanted upshifts or downshifts that could compromise the vehicle's balance or traction. "Manumatic" shifters, first popularized by Porsche in the 1990s under the trade name Tiptronic, have become a popular option on sports cars and other performance vehicles. With the near-universal prevalence of electronically controlled transmissions, they are comparatively simple and inexpensive, requiring only software changes, and the provision of the actual manual controls for the driver. The amount of true manual control provided is highly variable: some systems will override the driver's selections under certain conditions, generally in the interest of preventing engine damage. Since these gearboxes also have a throttle kickdown switch, it is impossible to fully exploit the engine power at low to medium engine speeds.

Second Gear Takeoff: Some automatics, particularly those fitted to larger capacity or high torque engines, either when "2" is manually selected, or by engaging a winter mode, will start off in second gear instead of first, and then not shift into a higher gear until returned to "D." Also note that as with most American automatic transmissions, selecting "2" using the selection lever will not tell the transmission to be in only 2nd gear; rather, it will simply limit the transmission to 2nd gear after prolonging the duration of 1st gear through higher speeds than normal operation. The 2000–2002 Lincoln LS V8 (the five-speed automatic *without* manumatic capabilities, as opposed to the optional sport package w/ manu-matic 5-speed) started in 2nd gear during most starts both in winter and other seasons by selecting the "D5" transmission selection notch in the shiftgate (for fuel savings), whereas "D4" would always start in 1st gear. This is done to reduce torque multiplication when proceeding forward from a standstill in conditions where traction was limited — on snow- or ice-covered roads, for example.

Some automatic transmissions modified or designed specifically for drag racing may also incorporate a transmission brake, or "trans-brake," as part of a manual valve body. Activated by electrical solenoid control, a trans-brake simultaneously engages the first and reverse gears, locking the transmission and preventing the input shaft from turning. This allows the driver of the car to raise the engine RPM against the resistance of the torque converter, then launch the car by simply releasing the trans-brake switch.

Hydraulic Brake

The hydraulic brake is an arrangement of braking mechanism which uses brake fluid, typically containing ethylene glycol, to transfer pressure from the controlling mechanism to the braking mechanism.

History

Fred Duesenberg originated hydraulic brakes on his 1914 racing cars and Duesenberg was the first automotive marque to use the technology on a passenger car in 1920. This braking system could have earned him a fortune if he had patented it. In 1918 Malcolm Lougheed (who later changed the spelling of his name to Lockheed) developed a hydraulic brake system.

Construction

The most common arrangement of hydraulic brakes for passenger vehicles, motorcycles, scooters, and mopeds, consists of the following:

- Brake pedal or lever
- A pushrod (also called an *actuating rod*)
- A master cylinder assembly containing a piston assembly (made up of either one or two pistons, a return spring, a series of gaskets/ O-rings and a fluid reservoir)
- Reinforced hydraulic lines
- Brake caliper assembly usually consisting of one or two hollow aluminium or chrome-plated steel pistons (called *caliper pistons*), a set of thermally conductive brake pads and a rotor (also called a *brake disc*) or drum attached to an axle.

The system is usually filled with a glycol-ether based brake fluid (other fluids may also be used).

At one time, passenger vehicles commonly employed drum brakes on all four wheels. Later, disc brakes were used for the front and drum brakes for the rear. However disc brakes have shown better heat dissipation and greater resistance to 'fading' and are therefore generally safer than drum brakes. So four-wheel disc brakes have become increasingly popular, replacing drums on all but the most basic vehicles. Many two-wheel vehicle designs, however, continue to employ a drum brake for the rear wheel.

The following description uses the terminology for and configuration of a simple disc brake.

System Operation

In a hydraulic brake system, when the brake pedal is pressed, a pushrod exerts force on the piston(s) in the master cylinder, causing fluid from the brake fluid reservoir to flow into a pressure chamber through a compensating port. This results in an increase in the pressure of the entire hydraulic system. This forces fluid through the hydraulic lines toward one or more calipers where it acts upon one or two caliper pistons sealed by one or more seated O-rings which prevent the escape of any fluid from around the piston.

The brake caliper pistons then apply force to the brake pads, which pushes them against the spinning rotor, and the friction between the pads and the rotor causes a braking torque to be generated, slowing the vehicle. Heat generated by this friction is either dissipated through vents and channels in the rotor or conducted through the pads, which are made of specialised heat-tolerant materials such as kevlar or sintered glass.

Subsequent release of the brake pedal/lever allows spring(s) to return the master piston(s) back into position. This relieves the hydraulic pressure on the caliper, allowing the brake piston in the caliper assembly to slide back into its housing and the brake pads to release the rotor.

The hydraulic braking system is designed as a closed system: unless there is a leak in the system, none of the brake fluid enters or leaves it, nor does the fluid get consumed through use.

An Example of a Hydraulic Brake System

Hydraulic brakes transfer energy to stop an object, normally a rotating axle. In a very simple brake system, with just two cylinders and a disc brake, the cylinders could be connected via tubes, with a piston inside the cylinders. The cylinders and tubes are filled with incompressible oil. The two cylinders have the same volume, but different diameters, and thus different cross-section areas. The one with the smallest diameter is called the master cylinder. The spinning disc brake will be placed down at the piston with the larger cross-section. Suppose the diameter of the master cylinder is half the diameter of the slave cylinder, so the master cylinder has a cross-section four times smaller. Now, if the piston in the master cylinder is pushed down 40 mm, with 10 newtons (N) of force, the slave piston will then move 10 mm, with a force of 40 N.

This force can be further increased by inserting a lever connected between the master piston, a pedal, and a pivot point. If the distance from the pedal to the pivot is three times the distance from the pivot to the connected piston, then it multiplies the pedal force by a factor of 3, when pushing down on the pedal. Now, if we push the pedal 120 mm down, with 10 N of force, a force of 30 N will then be applied to the master piston, and the slave piston will move the brake pad by 10 mm, with a force of 120 N.

Component Specifics

The brake pedal is a simple lever. One end is attached to the framework of the vehicle, a pushrod extends from a point along its length, and the foot pad is at the other end of the lever. The rod either extends to the master cylinder (manual brakes) or to the vacuum booster (power brakes).

In a four-wheel car, the FMVSS Standard 105, 1976; requires the master cylinder is divided internally into two sections, each of which pressurizes a separate hydraulic circuit. Each section supplies pressure to one circuit. Passenger vehicles typically have either a front/rear split brake system or a diagonal split brake system (the master cylinder in a motorcycle or scooter may only pressurize a single unit, which will be the front brake).

A front/rear split system uses one master cylinder section to pressurize the front caliper pistons and the other section to pressurize the rear caliper pistons. A split circuit braking system is now required by law in most countries for safety reasons; if one circuit fails, the other circuit can stop the vehicle.

Diagonal split systems were used initially on American Motors automobiles in the 1967 production year. The right front and left rear are served by one actuating piston while the left front and the right rear are served, exclusively, by a second actuating piston (both pistons pressurize their respective coupled lines from a single foot pedal). If either circuit fails, the other, with at least one front wheel braking (the front brakes provide most of the speed reduction) remains intact to stop the mechanically-damaged vehicle. Just before 1970, diagonally split systems had become universal for automobiles sold in the United States. This system was developed with front wheel drive cars suspension design to maintain better control and stability during a system failure.

The diameter and length of the master cylinder has a significant effect on the performance of the brake system. A larger diameter master cylinder delivers more hydraulic fluid to the caliper pistons, yet requires more brake pedal force and less brake pedal stroke to achieve a given deceleration. A smaller diameter master cylinder has the opposite effect. A master cylinder may also use differing diameters between the two sections to allow for increased fluid volume to one set of caliper pistons or the other.

Power Brakes

The vacuum booster or vacuum servo is used in most modern hydraulic brake systems which contain four wheels. The vacuum booster is attached between the master cylinder and the brake pedal and multiplies the braking force applied by the driver. These units consist of a hollow housing with a movable rubber diaphragm across the centre, creating two chambers. When attached to the low-pressure portion of the throttle body or intake manifold of the engine, the pressure in both chambers of the unit is lowered. The equilibrium created by the low pressure in both chambers keeps the diaphragm from moving until the brake pedal is depressed. A return spring keeps the diaphragm in the starting position until the brake pedal is applied. When the brake pedal is applied, the movement opens an air valve which lets in atmospheric pressure air to one chamber of the booster. Since the pressure becomes higher in one chamber, the diaphragm moves toward the lower pressure chamber with a force created by the area of the diaphragm and the differential pressure.

This force, in addition to the driver's foot force, pushes on the master cylinder piston. A relatively small diameter booster unit is required; for a very conservative 50% manifold vacuum, an assisting force of about 1500 N (200n) is produced by a 20 cm diaphragm with an area of 0.03 square metres. The diaphragm will stop moving when the forces on both sides of the chamber reach equilibrium. This can be caused by either the air valve closing (due to the pedal apply stopping) or if "run out" is reached. Run out occurs when the pressure in one chamber reaches atmospheric pressure and no additional force can be generated by the now stagnant differential pressure. After the run out point is reached, only the driver's foot force can be used to further apply the master cylinder piston.

The fluid pressure from the master cylinder travels through a pair of steel brake tubes to a pressure differential valve, sometimes referred to as a "brake failure valve", which performs two functions: it equalizes pressure between the two systems, and it provides a warning if one system loses pressure. The pressure differential valve has two chambers (to which the hydraulic lines attach) with a piston between them. When the pressure in either line is balanced, the piston does not move. If the pressure on one side is lost, the pressure from the other side moves the piston. When the piston makes contact with a simple electrical probe in the centre of the unit, a circuit is completed, and the operator is warned of a failure in the brake system.

From the pressure differential valve, brake tubing carries the pressure to the brake units at the wheels. Since the wheels do not maintain a fixed relation to the automobile, it is necessary to use hydraulic brake hose from the end of the steel line at the vehicle frame to the caliper at the wheel. Allowing steel brake tubing to flex invites metal fatigue and, ultimately, brake failure. A common upgrade is to replace the standard rubber hoses with a set which are externally reinforced with braided stainless-steel wires; these have negligible expansion under pressure and can give a firmer feel to the brake pedal with less pedal travel for a given braking effort.

Special Considerations

Air brake systems are bulky, and require air compressors and reservoir tanks. Hydraulic systems are smaller and less expensive.

Hydraulic fluid must be non-compressible. Unlike air brakes, where a valve is opened and air flows into the lines and brake chambers until the pressure rises sufficiently, hydraulic systems rely on a single stroke of a piston to force fluid through the system. If any vapor is introduced into the system it will compress, and the pressure may not rise sufficiently to actuate the brakes. Hydraulic braking systems are sometimes subjected to high temperatures during operation, such as when descending steep grades. For this reason, hydraulic fluid must resist vaporization at high temperatures.

Water vaporizes easily with heat and can corrode the metal parts of the system. If it gets into the brake lines, it can degrade brake performance dramatically. This is why light oils are often used as hydraulic fluids. Oil displaces water, protects plastic parts against corrosion, and can tolerate much higher temperatures before vaporizing.

"Brake fade" is a condition caused by overheating in which braking effectiveness reduces, and may be lost. It may occur for many reasons. The pads which engage the rotating part may become overheated and "glaze over", becoming so smooth and hard that they cannot grip sufficiently to slow the vehicle. Also, vaporization of the hydraulic fluid under temperature extremes or thermal distortion may cause the linings to change their shape and engage less surface area of the rotating part. Thermal distortion may also cause permanent changes in the shape of the metal components, resulting in a reduction in braking capability that requires replacement of the affected parts.

High-Density Solids Pump

High-density solids pumps are hydrostatically operating machines which displace the medium being pumped and thus create a flow.

High-Density Solids and Their Transport

High-density solids are mixtures of liquid and solid constituents. Each have their own typical physical and chemical characteristics such as specific weight, solids content, maximum particle size as well as how they behave, for example thixotropical, abrasively or adhesively. Typical examples for high-density solids are sludges and slurries.

Within industrial plants many substances with high proportion of solids - such as mechanically dewatered sewage sludges, filter cakes, bituminous coal sludges, waste and process sludges - have to be transported over relatively great distances to landfills or incineration plants.

In principle, long conveying distances can be bridged with mechanical conveying equipment or hydraulical high-density solids pumps.

Types of mechanical conveyors are belt, screw (worm), or scraper. They are suitable for nearly all types of high-density solids.

Hydraulical high-density solids pumps move sludges and slurries within an enclosed pipeline. However not every sludge or slurry is pumpable, depending on:

- Ratio: The mixture of solid and liquid components must be such that it results in a plastically deformable mass.
- Saturation: The volume of pore space in the sludges must be filled with sufficient quantity of solid matter for particle to be able to rest on particle via a plastically viscous liquid and for interstices to be filled.
- Gas component: As a consequence of pressure fluctuations knocking in the pipe may happen. By injection of about 2% air, sludges which do not naturally have a gas component can be brought to a better consistency.

Types of High-Density Solids Pumps

Depending on how the displacement principle is implemented, a distinction can be drawn between rotary and reciprocating pumps. The rotary circulation pumps includes eccentric screw pumps, centrifugal pumps, and squeezed tube (peristaltic) pumps. Reciprocating pumps include plunger, diaphragm, and piston pumps.

The high-density solids pumps which can be used in the widest range of situations is the hydraulically powered piston (or reciprocating) pump. It can be realised as single-cylinder or two-cylinder pump. In the latter case, power is transferred to the material via delivery pistons working in push-pull-mode. While one delivery piston sucks material from the feed hopper into the cylinder, the second piston simultaneously pushes the material in the other delivery cylinder and into the delivery line.

Type: Piston Pump with Transfer Tube (Two-Cylinder Pump)

Construction Details: The most characteristic element of the transfer tube pump is the so-called "S" transfer tube installed inside the feed hopper. The job of the transfer tube is to control the flow of high-density solids - in other words, the pump cylinder which is currently pushing the material is connected to the delivery line by such tube. At (or nearby) the end of each piston stroke the S-shaped tube which is connected to the delivery line is swung over to the other delivery cylinder with the aid of two switch cylinders. A hydraulic circuit is used for synchronizing the position of the transfer tube and the movements of the two delivery pistons.

The S tube should swing over rapidly and not suffer undue wear. This is effected with two large-sized plunger cylinders and a so-called spectacle plate (so named from its shape) including seal rings.

In cross section the S-tube is circular with a diameter which tapers in the direction of flow. This keeps the risk of clogging down. The sealing faces of the spectacle plate and of the seal ring lie parallel with the swivelling movement of the S-tube, so that they cannot be damaged when foreign bodies are cut through or get stuck.

This pump does not need check valves.

Operating Features

Piston pumps with S-tube are suitable for concrete and mortar, for sludges with a solids content of up to 50% by weight and a fluctuating particle size distribution but also for conveying fly ash, coal or minerals in a suspension.

The large cross section of the holes in the spectacle plate means that with a delivery cylinder diameter of e.g. 200 mm (7.874 in) it is possible to pump high-density solids with a mean grain size of as much as 80 mm (3.150 in). The maximum diameter of individual foreign bodies may be as high as 60% of the diameter of the delivery line - in this example this would be 120 mm (4.724 in).

Type: Piston Pump with Seat Valves (Two-Cylinder Pump)

Construction details: The steel housing of the seat valve pump contains four hydraulically controlled seat valves, two suction and two delivery valves for the two hydraulic cylinders. The suction and discharge valves are synchronized with the hydraulics of the delivery pistons - this ensures that the contents of the delivery cylinder is equal at all times to the volume being pumped. Once the "sucking" piston reaches its end position, the corresponding suction or delivery valve is simultaneously closed or opened. If there is overpressure in the delivery line, the suction valve will close first. This prevents the pumped medium from being pushed out of the pressure line back into the hopper.

This pump does not need check valves.

Operating Features

Seat valve pumps are suitable for an especially even pumping of media with a solids content of up to 50% and for the high-pressure pumping of paste-like industrial media such as sludges. Since valves are used in this two-cylinder pump it is suitable for grain sizes up to 8mm at most.

Which type of valve is selected will depend on the high-density solids to be conveyed. For materials containing grains with a solids content up to 50% a sharp-edged metal valve seat is required. With low-viscosity, rather watery fine sludges the choice would be a large-area elastomeric valve seal.

In one hydraulic control variant the valve is opened passively by the thrust of the medium. In this case the discharge valve has a non-return function which prevents flowback from the pressure line. At the same time the high-density solids are precompacted to close to the line pressure before the discharge valve opens. This means that pipe knocking resulting from pressure pulsations can be avoided.

Type: Ball-Valve Pump (Two-Cylinder Pump)

Construction Details: During the suction stroke of the delivery piston the medium is sucked in via the open suction ball valve. At the same time the ball valve on the delivery line is pushed into the valve seating by the underpressure resulting from the suction (automatic valves). In parallel with this, the second delivery piston is making the pressure stroke and forces the medium through the discharge ball valve and into the delivery line. The suction ball valve is pushed into its seating by the delivery line pressure and thereby closes off the connection to pump intake.

Operating Features

The two-cylinder ball valve pump is suitable for in the low to middle pressure range for pumping highly liquid to paste-like media provided these can be sucked in through the valve openings - for example mortars, mineral and sewage sludges.

The ball valve pump is to all intents and purposes insensitive to corrosive and abrasive media since the ball valves do not have to be opened or closed from the outside.

Type: Single-cylinder piston pump

Construction Details: With a single-cylinder piston pump the medium is "stuffed" out of a feed chute into the delivery cylinder and by the cylinder into the delivery line on the forward stroke. Depending on the level of the delivery pressure in the pipe and on the flow properties of the medium, the delivery flange of the pump will be fitted with a flat slide valve or a lamellar non-return valve in order to prevent flowback of the material on the return stroke. Which delivery piston is actually selected will depend on the material which is to be pumped: Free flowing media call for perturbane sealing elements while bulky materials which have to be chopped up during pumping mean that hardened cutting edges are required.

Operating Features

The single-cylinder pump is suitable for bulk materials with an awkward coarse composition - for example wood chips, organic waste, hazardous waste, or shredded paper. If such unsaturated material is to be conveyed over large distances, one possible approach is the hybrid feed system. Here compressed air is injected into the delivery line. It presses the plug of material into the pipe. As the compressed air expands the material being conveyed is loosened and separated. As distance increases from the air injection point, what was initially a series of plugs gradually turns into a continuous airborne stream. The amount of air required and the appropriate pipe diameter depend on the type and quantity of matter being conveyed.

Parts of High-Density Solids Piston Pumps

Main parts are:

- feed hopper
- drive cylinder, delivery cylinder and hydraulic system
- seals
- electrical and control equipment

  and - depending on the type:
- transfer tube
- valves

  as well as peripheral components such as:

- bunker (with / without sliding frames)
- screw conveyors, pre-press equipment
- pipework
- air injection
- boundary layer injection
- systems for pressure (pulsation) damping

Solids Pump Selection

The special kind of medium "high-density solids" and the wide range of applications require different construction principles and hence pump types. Following table permits a pre-selection depending on the media characteristics:

Type	*Medium (High-Density Solids)*			
	Watery	*Stiff*	*Bulky*	*with Foreign Bodies*
Transfer type pump	+	+ +	-	+
Seat valve pump	+ +	+	-	-
Ball valve pump	+	o	-	-
Single-cylinder pump	-	+ +	+ +	+ +

+ + very suitable, + suitable, o with reservations, - not suitable

Applications

The hydraulic pumping of solids with high solids contents offers a range of possibilities in process engineering. Oil-hydraulically powered two-cylinder piston pumps with S transfer tube have already been used for many years in civil engineering for transporting concrete. This technology has been transferred to further areas of application in various sectors of industry.

Sewage Treatment Plants

Mechanically dewatered sewage sludge can be pumped directly into furnaces via a pipe network. This avoids direct contact with the sludges and prevents from contamination and odour emission.

Waste Management (Hazardous Waste)

Hazardous and highly flammable substances are processed physically, chemically and biologically in order to reduce the waste materials. With the advantage of pipeline conveyance, high density solids pumps commonly transport such material.

Mining

Real tunnelling (in other words, not cut-and-cover) has increased palpably in recent years. Conveying mine water, soil and tailings to the tunnel entrance is a need.

Boundary Conditions

Transport Routes: Various aspects have to be considered in the planning and processing of conveying installations. This includes the space requirements, the accessibility in case of maintenance or cleaning purposes, as well as horizontal or vertical transport route for material discharge.

Environmental Protection (Noise and Odour Reduction)

Due to the nature of the material and the conveying method, odour nuisances is to be expected. Choosing the appropriate conveying method can reduce the environmental pollution.

Energy Expenditure

Three forces have to be overcome when conveying with high density solids pumps:

1. The conveying distance (cf. Record held by Concrete pump)
2. Friction between material and pipeline, especially in bends and reduction joints.
3. Inner frictional force caused by deformation of the material.

Calculation example:

- Conveyance suspension
- Delivery line of 200 mm (8 in) diameter (DN200)
- High-density solids with mineral or carbon content as a mixture from coarse and fine grained material.
- max. particle size 50 mm (1.969 in)
- Dry solids content up to 80 percent by weight

Energy consumption = approx. 50 Wh (180 kJ) per tonne of conveyed material and per kilometre (280 Btu per long ton or 250 Btu per short ton per mile).

Chapter 4

Hydraulic Accumulator

A hydraulic accumulator is a pressure storage reservoir in which a non-compressible hydraulic fluid is held under pressure by an external source. The external source can be a spring, a raised weight, or a compressed gas. An accumulator enables a hydraulic system to cope with extremes of demand using a less powerful pump, to respond more quickly to a temporary demand, and to smooth out pulsations.

It is a type of energy storage device. Compressed gas accumulators, also called hydro-pneumatic accumulators, are by far the most common type.

Types of Accumulators

Towers: The first accumulators for Armstrong's hydraulic dock machinery were simple raised water towers. Water was pumped to a tank at the top of these towers by steam pumps. When dock machinery required hydraulic power, the hydrostatic head of the water's height above ground provided the necessary pressure. These simple towers were extremely tall.

One of the best known, Grimsby Dock Tower opened in 1852, is 300 feet (91 m) tall. The size of these towers made them expensive to construct. By the time Grimsby was opened, it was already obsolete as Armstrong had developed the more complex, but much smaller, weighted accumulator.

These simple tower accumulators were constructed for less than a decade. In 1892 the original Grimsby tower's function was replaced by a smaller weighted accumulator on an adjacent dock, although the tower remains to this day as a well-known landmark.

Wallasey Great Float

Raised Weight

A raised weight accumulator consists of a vertical cylinder containing fluid connected to the hydraulic line. The cylinder is closed by a piston on which a series of weights are placed that exert a downward force on the piston and thereby energizes the fluid in the cylinder. In contrast to compressed gas and spring accumulators, this type delivers a nearly constant pressure, regardless of the volume of fluid in the cylinder, until it is empty. (The pressure will decline somewhat as the cylinder is emptied due to the decline in weight of the remaining fluid.)

A working example of this type of accumulator may be found at the hydraulic engine house, Bristol Harbour. The external accumulator was added around 1920. The water is pumped from the harbour into a header tank and then fed by gravity to the pumps. The working pressure is 750 psi (5.2 MPa, or 52 bar) which is used to power the cranes, bridges and locks of Bristol Harbour.

The original operating mechanism of Tower Bridge, London, also used this type of accumulator. Although no longer in use, two of the six accumulators may still be seen *in situ* in the bridge's museum.

Regent's Canal Dock, now named Limehouse Basin has the remains of a hydraulic accumulator, dating from 1869, a fragment of the oldest remaining such facility in the world, the second at the dock, which was installed later than that at Poplar Dock, originally listed incorrectly as a signalling cabin for the London and Blackwall Railway, when correctly identified, it was restored as a tourist attraction by the now defunct London Docklands Development Corporation. Now owned by the British Waterways Board, it is open for large groups on application to the Dockmaster's Office at the basin and on both the afternoons of London Open House Weekend, held on the third weekend of September each year.

London had an extensive public hydraulic power system from the mid-nineteenth century finally closing in the 1970s with 5 hydraulic power stations, operated by the London Hydraulic Power Company. Railway goods yards and docks often had their own separate system.

Compressed Gas (or Gas-Charged) Closed Accumulator

A compressed gas accumulator consists of a cylinder with two chambers that are separated by an elastic diaphragm, a totally enclosed bladder, or a floating piston. One chamber contains hydraulic fluid and is connected to the hydraulic line. The other chamber contains an inert gas under pressure (typically nitrogen) that provides the compressive force on the hydraulic fluid. Inert gas is used because oxygen and oil can form an explosive mixture when combined under high pressure. As the volume of the compressed gas changes, the pressure of the gas (and the pressure on the fluid) changes inversely. Existing hydraulic accumulator designs are large and heavy due to the need for two storage tanks and do not have the high energy density needed for many applications.

It is possible to increase the gas volume of the accumulator by coupling a gas bottle to the gas side of the accumulator. This is mainly done since a gas bottle normally is cheaper to produce than an accumulator. The compressed gas accumulator was invented by Jean Mercier, for use in variable pitch propellers.

Compressed Gas Open Accumulator

Hydraulic energy storage systems store energy by compressing air similar to a battery storing energy in an electric circuit. The need for two storage tanks and two accumulators can be eliminated and the entire hydraulic energy storage system is an open loop. The storage requirement is smaller because depressurized air is not stored. The hydraulic open loop accumulator works by drawing air in from the atmosphere and expelling air into the atmosphere. A separate hydraulic pump maintains the pressure balance of the air by increasing the amount of hydraulic fluid in the system. This results in a steady pressure of air and up to 24 times the energy density of a standard hydraulic accumulator. This hydraulic energy storage system has applications in energy storage for wind turbines, regenerative braking systems for hybrid cars that could partially power the car, and energy storage for power construction equipment.

Spring Type

A spring type accumulator is similar in operation to the gas-charged accumulator above, except that a heavy spring (or springs) is used to provide the compressive force. According to Hooke's law the magnitude of the force exerted by a spring is linearly proportional to its extension. Therefore as the spring compresses, the force it exerts on the fluid is increased linearly.

Metal Bellows Type

The metal bellows accumulators function similarly to the compressed gas type, except the elastic diaphragm or floating piston is replaced by a hermetically sealed welded metal bellows. Fluid may be internal or external to the bellows. The advantages to the metal bellows type include exceptionally low spring rate, allowing the gas charge to do all the work with little change in pressure from full to empty, and a long stroke relative to solid (empty) height, which gives maximum storage volume for a given container size. The welded metal bellows accumulator provides an exceptionally high level of accumulator performance, and can be produced with a broad spectrum of alloys resulting in a broad range of fluid compatibility. Another advantage to this type is that it does not face issues with high pressure operation, thus allowing more energy storage capacity.

Functioning of An Accumulator

In modern, often mobile, hydraulic systems the preferred item is a gas charged accumulator, but simple systems may be spring-loaded. There may be more than one accumulator in a system. The exact type and placement of each may be a compromise due to its effects and the costs of manufacture. An accumulator is placed close to the pump with a non-return valve preventing flow back to the pump. In the case of piston-type pumps this accumulator is placed in the ideal location to absorb pulsations of energy from the multi-piston pump. It also helps protect the system from fluid hammer. This protects system components, particularly pipework, from both potentially destructive forces.

An additional benefit is the additional energy that can be stored while the pump is subject to low demand. The designer can use a smaller-capacity pump. The large excursions of system components, such as landing gear on a large aircraft, that require a considerable volume of fluid can also benefit from one or more accumulators. These are often placed close to the demand to help overcome restrictions and drag from long pipework runs. The outflow of energy from a discharging accumulator is much greater, for a short time, than even large pumps could generate. An accumulator can maintain the pressure in a system for periods when there are slight leaks without the pump being cycled on and off constantly. When temperature changes cause pressure excursions the accumulator helps absorb them. Its size helps absorb fluid that might otherwise be locked in a small fixed system with no room for expansion due to valve arrangement.

The gas precharge in an accumulator is set so that the separating bladder, diaphragm or piston does not reach or strike either end of the operating cylinder. The design precharge normally ensures that the moving parts do not foul the ends or block fluid passages. Poor maintenance of precharge can destroy an operating accumulator. A properly designed and maintained accumulator should operate trouble-free for years.

Hydraulic Cylinder

A Hydraulic cylinder (also called a linear hydraulic motor) is a mechanical actuator that is used to give a unidirectional force through a unidirectional stroke. It has many applications, notably in construction equipment (engineering vehicles), manufacturing machinery, and civil engineering.

Operation

Hydraulic cylinders get their power from pressurized hydraulic fluid, which is typically oil. The hydraulic cylinder consists of a cylinder barrel, in which a piston connected to a piston rod moves back and forth. The barrel is closed on one end by the cylinder bottom (also called the cap) and the other end by the cylinder head (also called the gland) where the piston rod comes out of the cylinder. The piston has sliding rings and seals. The piston divides the inside of the cylinder into two chambers, the bottom chamber (cap end) and the piston rod side chamber (rod end / head end).

Flanges, trunnions, clevises, Lugs are common cylinder mounting options. The piston rod also has mounting attachments to connect the cylinder to the object or machine component that it is pushing / pulling. A hydraulic cylinder is the actuator or "motor" side of this system. The "generator" side of the hydraulic system is the hydraulic pump which brings in a fixed or regulated flow of oil to the hydraulic cylinder, to move the piston. The piston pushes the oil in the other chamber back to the reservoir. If we assume that the oil enters from cap end, during extension stroke, and the oil pressure in the rod end / head end is approximately zero, the force F on the piston rod equals the pressure P in the cylinder times the piston area A:

$$F = P \cdot A$$

During the retraction stroke, if oil is pumped into the head (or gland) at the rod end and the oil from the cap end flows back to the reservoir without pressure, the fluid pressure in the rod end is (Pull Force) / (piston area - piston rod area):

$$P = \frac{F_p}{A_p - A_r}$$

where P is the fluid pressure, F_p is the pulling force, A_p is the piston face area and A_r is the rod cross-section area.

Parts of a Hydraulic Cylinder

A hydraulic cylinder consists of the following parts:

Cylinder Barrel

The main function of cylinder body is to hold cylinder pressure. The cylinder barrel is mostly made from a seamless tube. The cylinder barrel is ground and/or honed internally with a typical surface finish of 4 to 16 microinch. Normally hoop stress is calculated to optimize the barrel size.

Cylinder Base or Cap

The main function of the cap is to enclose the pressure chamber at one end. The cap is connected to the body by means of welding, threading, bolts, or tie rod. Caps also perform as cylinder mounting components [cap flange, cap trunnion, cap clevis]. Cap size is determined based on the bending stress. A static seal / o-ring is used in between cap and barrel (except welded construction).

Cylinder Head

The main function of the head is to enclose the pressure chamber from the other end. The head contains an integrated rod sealing arrangement or the option to accept a seal gland. The head is connected to the body by means of threading, bolts, or tie rod. A static seal / o-ring is used in between head and barrel.

Piston

The main function of the piston is to separate the pressure zones inside the barrel. The piston is machined with grooves to fit elastomeric or metal seals and bearing elements. These seals can be single acting or double acting. The difference in pressure between the two sides of the piston causes the cylinder to extend and retract. The piston is attached with the piston rod by means of threads, bolts, or nuts to transfer the linear motion.

Piston Rod

The piston rod is typically a hard chrome-plated piece of cold-rolled steel which attaches to the piston and extends from the cylinder through the rod-end head. In double rod-end cylinders, the actuator has a rod extending from both sides of the piston and out both ends of the barrel. The piston rod connects the hydraulic actuator to the machine component doing the work. This connection can be in the form of a machine thread or a mounting attachment....

Seal Gland

The cylinder head is fitted with seals to prevent the pressurized oil from leaking past the interface between the rod and the head. This area is called the seal gland. The advantage of a seal gland is easy removal and seal replacement. The seal gland contains a primary seal, a secondary seal / buffer seal, bearing elements, wiper / scraper and static seal. In some cases, especially in small hydraulic cylinders, the rod gland and the bearing elements are made from a single integral machined part.

Seals

The seals are considered / designed as per the cylinder working pressure, cylinder speed, operating temperature, working medium and application. Piston seals are dynamic seals, and they can be single acting or double acting. Generally speaking, Elastomer seals made from nitrile rubber, Polyurethane or other materials are best in lower temperature environments, while seals made of Fluorocarbon Viton are better for higher temperatures. Metallic seals are also available and commonly use cast iron for the seal material. Rod seals are dynamic seals and generally are single acting. The compounds of rod seals are nitrile rubber, Polyurethane, or Fluorocarbon Viton. Wipers / scrapers are used to eliminate contaminants such as moisture, dirt, and dust, which can cause extensive damage to cylinder walls, rods, seals and other components. The common compound for wipers is polyurethane. Metallic scrapers are used for sub zero temperature applications, and applications where foreign materials can deposit on the rod. The bearing elements / wear bands are use to eliminate metal to metal contact. The wear bands are designed as per the side load requirements. The primary compounds for wear bands are filled PTFE, woven fabric reinforced polyester resin and bronze.

Other Parts

- Cylinder base connection
- Cushions

Single Acting vs. Double Acting

- Single acting cylinders are economical and the simplest design. Hydraulic fluid enters through a port at one end of the cylinder, which extends the rod by means of area difference. An external force or gravity returns the piston rod.

- Double acting cylinders have a port at each end, supplied with hydraulic fluid for both the retraction and extension.

Hydraulic Cylinder Designs

There are primarily two styles of hydraulic cylinder construction used in industry: Tie rod style cylinders and welded body style cylinders.

Tie Rod Cylinder

Tie rod style hydraulic cylinders use high strength threaded steel rods to hold the two end caps to the cylinder barrel. This method of construction is most often seen in industrial factory applications. Small bore cylinders usually have 4 tie rods, while large bore cylinders may require as many as 16 or 20 tie rods in order to retain the end caps under the tremendous forces produced. Tie rod style cylinders can be completely disassembled for service and repair.

The National Fluid Power Association (NFPA) has standardized the dimensions of hydraulic tie rod cylinders. This enables cylinders from different manufacturers to interchange within the same mountings.

Welded Body Cylinder

Welded body cylinders have no tie rods. The barrel is welded directly to the end caps. The ports are welded to the barrel. The front rod gland is usually threaded into or bolted to the cylinder barrel. This allows the piston rod assembly and the rod seals to be removed for service. Welded body cylinders have a number of advantages over tie rod style cylinders. Welded cylinders have a narrower body and often a shorter overall length enabling them to fit better into the tight confines of machinery. Welded cylinders do not suffer from failure due to tie rod stretch at high pressures and long strokes. The welded design also lends itself to customization. Special features are easily added to the cylinder body. These may include special ports, custom mounts, valve manifolds, and so on.

The smooth outer body of welded cylinders also enables the design of multi-stage telescopic cylinders.

Welded body hydraulic cylinders dominate the mobile hydraulic equipment market such as construction equipment (excavators, bulldozers, and road graders) and material handling equipment (forklift trucks, telehandlers, and lift-gates). They are also used in heavy industry such as cranes, oil rigs, and large off-road vehicles in above-ground mining.

Piston Rod Construction

The piston rod of an hydraulic cylinder operates both inside and outside the barrel, and consequently both in and out of the hydraulic fluid and surrounding atmosphere.

Coatings

Wear and corrosion resistant surfaces are desirable on the outer diameter of the piston rod. The surfaces are often applied using coating techniques such as Chrome (Nickel) Plating, Lunac 2+ duplex, Laser Cladding, PTA welding and Thermal Spraying. These coatings can be finished to the desirable surface roughness (Ra, Rz) where the seals show optimum performance. All these coating methods have their specific advantages and disadvantages. It is for this reason that coating experts play a crucial role in selecting the optimum surface treatment procedure for protecting Hydraulic Cylinders. Cylinders are used in different operational conditions and that makes it a challenge to find the right coating solution. In dredging there might be impact from stones or other parts, in salt water environments there are extreme corrosion attacks, in off-shore cylinders facing bending and impact in combination with salt water, and in the steel industry there are high temperatures involved, etc.... It is important to understand that currently there is no single coating solution which successfully combats all the specific operational wear conditions. Every single technique has its own benefits and disadvantages.

Length

Piston rods are generally available in lengths which are cut to suit the application. As the common rods have a soft or mild steel core, their ends can be welded or machined for a screw thread.

Special Hydraulic Cylinders

Telescopic Cylinder: The length of an hydraulic cylinder is the total of the stroke, the thickness of the piston, the thickness of bottom and head and the length of the connections. Often this length does not fit in the machine. In that case the piston rod is also used as a piston barrel and a second piston rod is used. These kinds of cylinders are called telescopic cylinders. If we call a normal rod cylinder single stage, telescopic cylinders are multi-stage units of two, three, four, five or more stages. In general telescopic cylinders are much more expensive than normal cylinders. Most telescopic cylinders are single acting (push). Double acting telescopic cylinders must be specially designed and manufactured.

Plunger Cylinder

An hydraulic cylinder without a piston or with a piston without seals is called a plunger cylinder. A plunger cylinder can only be used as a pushing cylinder; the maximum force is piston rod area multiplied by pressure. This means that a plunger cylinder in general has a relatively thick piston rod.

Differential Cylinder

A differential cylinder acts like a normal cylinder when pulling. If the cylinder however has to push, the oil from the piston rod side of the cylinder is not returned to the reservoir, but goes to the bottom side of the cylinder. In such a way, the cylinder goes much faster, but the maximum force the cylinder can give is like a plunger cylinder. A differential cylinder can be manufactured like a normal cylinder, and only a special control is added.

Position Sensing "Smart" Hydraulic Cylinder

Position sensing hydraulic cylinders eliminate the need for a hollow cylinder rod. Instead, an external sensing "bar" using Hall Effect technology senses the position of the cylinder's piston. This is accomplished by the placement of a permanent magnet within the piston. The magnet propagates a magnetic field through the steel wall of the cylinder, providing a locating signal to the sensor.

A Note about Popular Terminology

In the United States, popular usage refers to the whole assembly of cylinder, piston, and piston rod (or more) collectively as a "piston", which is incorrect. Instead, the piston is the short, cylindrical metal component that separates the two parts of the cylinder barrel internally.

Hydraulic Ram

A hydraulic ram, or hydram, is a cyclic water pump powered by hydropower. It takes in water at one "hydraulic head" (pressure) and flow rate, and outputs water at a higher hydraulic head and lower flow rate. The device uses the water hammer effect to develop pressure that allows a portion of the input water that powers the pump to be lifted to a point higher than where the water originally started. The hydraulic ram is sometimes used in remote areas, where there is both a source of low-head hydropower and a need for pumping water to a destination higher in elevation than the source. In this situation, the ram is often useful, since it requires no outside source of power other than the kinetic energy of flowing water.

History

In 1772, John Whitehurst of Cheshire, United Kingdom, invented a manually controlled precursor of the hydraulic ram called the "pulsation engine" and installed the first one at Oulton, Cheshire to raise water to a height of 4.9 metres (16 ft). In 1783, he installed another in Ireland. He did not patent it, and details are obscure, but it is known to have had an air vessel.

The first self-acting ram pump was invented by the Frenchman Joseph Michel Montgolfier (best known as a co-inventor of the hot air balloon) in 1796 for raising water in his paper mill at Voiron. His friend Matthew Boulton took out a British patent on his behalf in 1797. The sons of Montgolfier obtained an English patent for an improved version in 1816, and this was acquired, together with Whitehurst's design, in 1820 by Josiah Easton, a Somerset-born engineer who had just moved to London.

Figure: *Hydraulic ram, System Lambach now at Roscheider Hof Open Air Museum*

Easton's firm, inherited by his son James (1796–1871), grew during the nineteenth century to become one of the more important engineering manufacturers in the United Kingdom, with a large works at Erith, Kent. They specialised in water supply and sewerage systems world-wide, as well as land drainage projects. Eastons had a good business supplying rams for water supply purposes to large country houses, farms, and village communities. Some of their installations still survived as of 2004.

The firm closed in 1909, but the ram business was continued by James R Easton. In 1929, it was acquired by Green & Carter of Winchester, Hampshire, who were engaged in the manufacturing and installation of the well-known Vulcan and Vacher Rams. The first US patent was issued to J. Cerneau and S.S. Hallet in 1809. US interest in hydraulic rams picked up around 1840, as further patents were issued and domestic companies started offering rams for sale. Toward the end of the 19th century, interest waned as electricity and electric pumps became widely available.

By the end of the twentieth century interest in hydraulic rams has revived, due to the needs of sustainable technology in developing countries, and energy conservation in developed ones. A good example is AID Foundation International in the Philippines, who won an Ashden Award for their work developing ram pumps that could be easily maintained for use in remote villages. The hydraulic ram principle has been used in some proposals for exploiting wave power, one of which was discussed as long ago as 1931 by Hanns Günther in his book *In hundert Jahren.*

Some later ram designs in the UK called compound rams were designed to pump treated water using an untreated drive water source, which overcomes some of the problems of having drinking water sourced from an open stream.

Construction and Principle of Operation

A hydraulic ram has only two moving parts, a spring or weight loaded "waste" valve sometimes known as the "clack" valve and a "delivery" check valve, making it cheap to build, easy to maintain, and very reliable. In addition, there is a drive pipe supplying water from an elevated source, and a delivery pipe, taking a portion of the water that comes through the drive pipe to an elevation higher than the source.

Sequence of Operation

1. Inlet – drive pipe
2. Free flow at waste valve
3. Outlet – delivery pipe
4. Waste valve
5. Delivery check valve
6. Pressure vessel

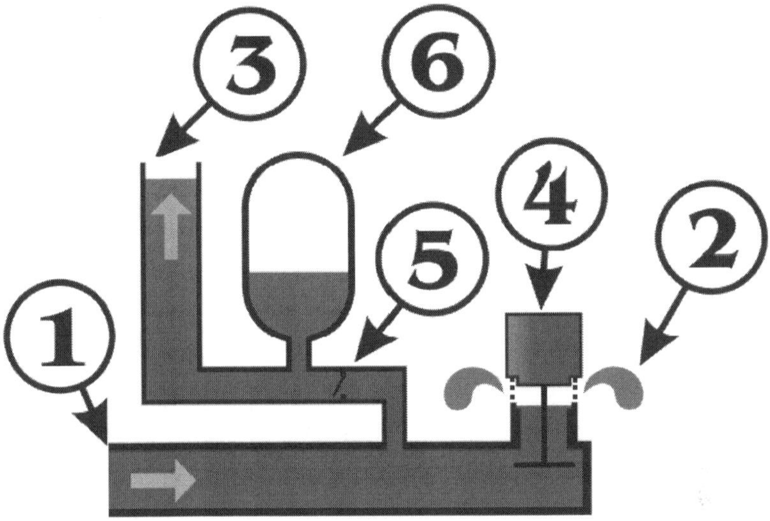

Figure: Basic components of a hydraulic ram:

A simplified hydraulic ram is shown in Figure depicted earlier. Initially, the waste valve [4] is open, and the delivery valve [5] is closed. The water in the drive pipe [1] starts to flow under the force of gravity and picks up speed and kinetic energy until the increasing drag force closes the waste valve. The momentum of the water flow in the supply pipe against the now closed waste valve causes a water hammer that raises the pressure in the pump, opens the delivery valve [5], and forces some water to flow into the delivery pipe [3]. Because this water is being forced uphill through the delivery pipe farther than it is falling downhill from the source, the flow slows; when the flow reverses, the delivery check valve closes. Meanwhile, the water hammer from the closing of the waste valve also produces a pressure pulse which propagates back up the supply pipe to the source where it converts to a suction pulse that propagates back down the pipe. This suction pulse, with the weight or spring on the valve, pulls the waste valve back open and allows the process to begin again.

A pressure vessel [6] containing air cushions the hydraulic pressure shock when the waste valve closes, and it also improves the pumping efficiency by allowing a more constant flow through the delivery pipe. Although, in theory, the pump could work without it, the efficiency would drop drastically and the pump would be subject to extraordinary stresses that could shorten its life considerably. One problem is that the pressurized air will gradually dissolve into the water until none remains. One solution to this problem is to have the air separated from the water by an elastic diaphragm (similar to an expansion tank); however, this solution can be problematic in developing countries where replacements are difficult to procure. Another solution is to have a mechanism such as a snifting valve that automatically inserts a small bubble of air when the suction pulse mentioned above reaches the pump. Another solution is to insert an inner tube of a car or bicycle tire into the pressure vessel with some air in it and the valve closed. This tube is in effect the same as the diaphragm, but it is implemented with more widely available materials. The air in the tube cushions the shock of the water the same as the air in other configurations does.

Efficiency

A typical energy efficiency is 60%, but up to 80% is possible. This should not be confused with the volumetric efficiency, which relates the volume of water delivered to total water taken from the source. The portion of water available at the delivery pipe will be reduced by the ratio of the delivery head to the supply head. Thus if the source is 2 metres above the ram and the water is lifted to 10 metres above the ram, only 20% of the supplied water can be available, the other 80% being spilled via the waste valve. These ratios assume 100% energy efficiency. Actual water delivered will be further reduced by the energy efficiency factor. In the above example, if the energy efficiency is 70%, the water delivered will be 70% of 20%, i.e. 14%. Assuming a 2 to one supply head to delivery head ratio and 70% efficiency, the delivered water would be 70% of 50%, i.e. 35%. Very high ratios of delivery to supply head usually result in lowered energy efficiency. Suppliers of rams often provide tables giving expected volume ratios based on actual tests.

Drive and Delivery Pipe Design

Since both efficiency and reliable cycling depend on water hammer effects, the drive pipe design is important. It should be between 3 and 7 times longer than the vertical distance between the source and the ram. Commercial rams may have an input fitting designed to accommodate this optimum slope. The diameter of the supply pipe would normally match the diameter of the input fitting on the ram, which in turn is based on its pumping capacity. The drive pipe should be of constant diameter and material, and should be as straight as possible. Where bends are necessary, they should be smooth, large diameter curves. Even a large spiral is allowed, but elbows are to be avoided. PVC will work in some installations, but steel pipe is preferred, although much more expensive. If valves are used they should be a free flow type such as a ball valve or gate valve. The delivery pipe is much less critical since the pressure vessel prevents water hammer effects from travelling up it. Its overall design would be determined by the allowable pressure drop based on the expected flow. Typically the pipe size will be about half that of the supply pipe, but for very long runs a larger size may be indicated. PVC pipe and any necessary valves are not a problem.

Starting Operation

A ram newly placed into operation or which has stopped cycling must be started as follows. If the waste valve is in the raised (closed) position, which is most common, it must be pushed down manually into the open position and released. If the flow is sufficient, it will then cycle at least once.

If it does not continue to cycle, it must be pushed down repeatedly until it cycles continuously on its own, usually after three or four manual cycles. If the ram stops with the waste valve in the down position it must be lifted manually and kept up for as long as necessary for the supply pipe to fill with water and for any air bubbles to travel up the pipe to the source. This may take a minute or more. Then it can be started manually by pushing it down a few times as described above.

Common Operational Problems

Failure to deliver sufficient water may be due to improper adjustment of the waste valve, having too little air in the pressure vessel, or simply attempting to raise the water higher than the level of which the ram is capable. The ram may be damaged by freezing in winter, or loss of air in the pressure vessel leading to excess stress on the ram parts. These failures will require welding or other repair methods and perhaps parts replacement.

It is not uncommon for an operating ram to require occasional restarts. The cycling may stop due to poor adjustment of the waste valve, or insufficient water flow at the source. Air can enter if the supply water level is not at least a few inches above the input end of the supply pipe. Other problems are blockage of the valves with debris, or improper installation, such as using a supply pipe of non uniform diameter or material, having sharp bends or a rough interior, or one that is too long or short for the drop, or is made of an insufficiently rigid material. A PVC supply pipe will work in some installations but is not as optimum as steel.

Water Powered Pump

An alternative to Hydraulic ram is water powered pump when the requirement is of high flow rate at high head ratio. Water powered pump unit is a hydraulic turbine coupled to a water pump. The motive power needed by the pump is generated by the hydraulic turbine from the available low head water energy.

Hydraulic Fluid

Hydraulic fluids, also called hydraulic liquids, are the medium by which power is transferred in hydraulic machinery. Common hydraulic fluids are based on mineral oil or water. Examples of equipment that might use hydraulic fluids include excavators and backhoes, hydraulic brakes, power steering systems, transmissions, garbage trucks, aircraft flight control systems, lifts, and industrial machinery. Hydraulic systems like the ones mentioned above will work most efficiently if the hydraulic fluid used has zero compressibility.

Functions and Properties

The primary function of a hydraulic fluid is to convey power. In use, however, there are other important functions of hydraulic fluid such as protection of the hydraulic machine components. The table below lists the major functions of a hydraulic fluid and the properties of a fluid that affect its ability to perform that function:

Function		*Property*
Medium for power	•	Non compressible (high bulk modulus)
transfer and control		
	•	Fast air release
	•	Low foaming tendency
	•	Low volatility
Medium for heat transfer	•	Good thermal capacity and conductivity
Sealing Medium	•	Adequate viscosity and viscosity index
	•	Shear stability
Lubricant	•	Viscosity for film maintenance · Low temperature fluidity · Thermal and oxidative stability · Hydrolytic stability / water tolerance · Cleanliness and filterability · Demulsibility

	• Antiwear characteristics
	• Corrosion control
Pump efficiency	• Proper viscosity to minimize internal leakage
	• High viscosity index
Special function	• Fire resistance
	• Friction modifications
	• Radiation resistance
Environmental impact	• Low toxicity when new or decomposed
	• Biodegradability
Functioning life	• Material compatibility

Composition

Base Stock: The original hydraulic fluid, dating back to the time of ancient Egypt, was water. Beginning in the 1920s, mineral oil began to be used more than water as a base stock due to its inherent lubrication properties and ability to be used at temperatures above the boiling point of water. Today most hydraulic fluids are based on mineral oil base stocks.

Natural oils such as rapeseed (also called canola oil) are used as base stocks for fluids where biodegradability and renewable sources are considered important.

Other base stocks are used for speciality applications, such as for fire resistance and extreme temperature applications. Some examples include: glycol, esters, organophosphate ester, polyalphaolefin, propylene glycol, and silicone oils.

Other Components

Hydraulic fluids can contain a wide range of chemical compounds, including: oils, butanol, esters (e.g. phthalates, like DEHP, and adipates, like bis(2-ethylhexyl) adipate), polyalkylene glycols (PAG), organophosphate (e.g. tributylphosphate), silicones, alkylated aromatic hydrocarbons, polyalphaolefins (PAO) (e.g. polyisobutenes), corrosion inhibitors (incl acid scavengers), anti-erosion additives, etc.

Biodegradable Hydraulic Fluids

Environmentally sensitive applications (e.g. farm tractors and marine dredging) may benefit from using biodegradable hydraulic fluids based upon rapeseed (Canola) vegetable oil when there is the risk of an oil spill from a ruptured oil line. Typically these oils are available as ISO 32, ISO 46, and ISO 68 specification oils. ASTM standards ASTM-D-6006, Guide for Assessing Biodegradability of Hydraulic Fluids and ASTM-D-6046, Standard Classification of Hydraulic Fluids for Environmental Impact are relevant.

Brake Fluid

Brake fluid is a subtype of hydraulic fluid with high boiling point, both when new (specified by the equilibrium boiling point) and after absorption of water vapor (specified by wet boiling point). Under the heat of braking, both free water and water vapor in a braking system can boil into a compressible vapor, resulting in brake failure. Glycol-ether based fluids are hygroscopic, and absorbed moisture will greatly reduce the boiling point over time. Silicone based fluids are not hygroscopic.

Safety

Because industrial hydraulic systems operate at hundreds to thousands of PSI and temperatures reaching hundreds of degrees Celsius, severe injuries and death can result from component failures and care must always be taken when performing maintenance on hydraulic systems. Fire resistance is a property available with specialised fluids.

Trade Names

Some of the trade names for hydraulic fluids include Arnica, Tellus, Durad, Fyrquel, Houghto-Safe, Hydraunycoil, Lubritherm Enviro-Safe, Pydraul, Quintolubric, Reofos, Reolube, Valvoline Ultramax and Skydrol.

Aircraft Hydraulic Systems

As aircraft performance increased in mid-20th century, the amount of force required to operate mechanical flight controls became excessive, and hydraulic systems were introduced to reduce pilot effort. The hydraulic actuators are controlled by valves; these in turn are operated directly by input from the aircrew (hydro-mechanical) or by computers obeying control laws (fly by wire). Hydraulic power is used for other purposes. It can be stored in accumulators to start an auxiliary power unit (APU) for self-starting the aircraft's main engines. Many aircraft equipped with the M61 family of cannon use hydraulic power to drive the gun system, permitting reliable high rates of fire.

The hydraulic power itself comes from pumps driven by the engines directly, or by electrically driven pumps. In modern commercial aircraft these are electrically driven pumps; should all the engines fail in flight the pilot will deploy a propeller-driven electric generator called a Ram Air Turbine (RAT) which is concealed under the fuselage. This provides electrical power for the hydraulic pumps and control systems as power is no longer available from the engines. In that system and others electric pumps can provide both redundancy and the means of operating hydraulic systems without the engines operating, which can be very useful during maintenance.

Specifications

Aircraft hydraulic fluids fall under various specifications:

Common petroleum-based:

- Mil-H-5606: Mineral base, flammable, fairly low flashpoint, usable from –65 °F (–54 °C) to 275 °F (135 °C), red colour
- Mil-H-83282: Synthetic hydrocarbon base, higher flashpoint, self-extinguishing, backward compatible to -5606, red colour, rated to –40 °F (–40 °C) degrees.
- Mil-H-87257: A development of -83282 fluid to improve its low temperature viscosity.

Phosphate-ester based:

- US/NATO Military Specification - MIL-H-8446
- Boeing Seattle - BMS3-11
- Boeing Long Beach - DMS2014
- Boeing Long Island - CDS5478
- Lockheed - LAC C-34-1224
- Airbus Industrie - NSA307110
- British Aerospace - BAC M.333.B
- Bombardier - BAMS 564-003
- SAE - Ac974
- SAE - AS1241

Contamination

Special, stringent care is required when handling aircraft hydraulic fluid as it is critical to flight safety that it stay free from contamination. It is also necessary to strictly adhere to authorized references when servicing or repairing any aircraft system. Samples from aircraft hydraulic systems are taken during heavy aircraft maintenance checks (primarily C and D checks) to check contamination.

Hydraulic Circuit

A hydraulic circuit is a system comprising an interconnected set of discrete components that transport liquid. The purpose of this system may be to control where fluid flows (as in a network of tubes of coolant in a thermodynamic system) or to control fluid pressure (as in hydraulic amplifiers).

The approach of describing a fluid system in terms of discrete components is inspired by the success of electrical circuit theory. Just as electric circuit theory works when elements are discrete and linear, hydraulic circuit theory works best when the elements (passive component such as pipes or transmission lines or active components such as power packs or pumps) are discrete and linear. This usually means that hydraulic circuit analysis works best for long, thin tubes with discrete pumps, as found in chemical process flow systems or microscale devices.

Components

The circuit comprises the following components:

- Active components
  - o Hydraulic power pack
- Transmission lines
  - o Hydraulic hoses
- Passive components
  - o Hydraulic cylinders

Hydraulic Motor

A hydraulic motor is a mechanical actuator that converts hydraulic pressure and flow into torque and angular displacement (rotation). The hydraulic motor is the rotary counterpart of the hydraulic cylinder.

Conceptually, a hydraulic motor should be interchangeable with a hydraulic pump because it performs the opposite function - much as the conceptual DC electric motor is interchangeable with a DC electrical generator. However, most hydraulic pumps cannot be used as hydraulic motors because they cannot be backdriven. Also, a hydraulic motor is usually designed for the working pressure at both sides of the motor.

Hydraulic pumps, motors, and cylinders can be combined into hydraulic drive systems. One or more hydraulic pumps, coupled to one or more hydraulic motors, constitutes a hydraulic transmission.

Figure: *Swing Bridge, River Tyne*

One of the first rotary hydraulic motors to be developed was that constructed by Armstrong for his Swing Bridge over the River Tyne. Two motors were provided, for reliability. Each one was a three-cylinder single-acting oscillating engine. Armstrong developed a wide range of hydraulic motors, linear and rotary, that were used for a wide range of industrial and civil engineering tasks, particularly for docks and moving bridges.

Gear and Vane Motors

Figure: *Hydraulic motor*

Gear and vane motors are used in simple rotating systems. Their benefits include low initial cost and high rpm.

A gear motor (external gear) consists of two gears, the driven gear (attached to the output shaft by way of a key, etc.) and the idler gear. High pressure oil is ported into one side of the gears, where it flows around the periphery of the gears, between the gear tips and the wall housings in which it resides, to the outlet port. The gears then mesh, not allowing the oil from the outlet side to flow back to the inlet side. For lubrication, the gear motor uses a small amount of oil from the pressurized side of the gears, bleeds this through the (typically) hydrodynamic bearings, and vents the same oil either to the low pressure side of the gears, or through a dedicated drain port on the motor housing. An especially positive attribute of the gear motor is that catastrophic breakdown is a lot less common than in most other types of hydraulic motors. This is because the gears gradually wear down the housing and/or main bushings, reducing the volumetric efficiency of the motor gradually until it is all but useless. This often happens long before wear causes the unit to seize or break down.

A vane motor consists of a housing with an eccentric bore, in which runs a rotor with vanes in it that slide in and out. The force differential created by the unbalanced force of the pressurized fluid on the vanes causes the rotor to spin in one direction. A critical element in vane motor design is how the vane tips are machined at the contact point between vane tip and motor housing. Several type of "lip" designs are used, and the main objective is to provide a tight seal between the inside of the motor housing and the vane, and at the same time to minimize wear and metal-to-metal contact.

Gerotor Motors

The gerotor motor is in essence a rotor with N-1 teeth, rotating off centre in a rotor/stator with N teeth. Pressurized fluid is guided into the assembly using a (usually) axially placed plate-type distributor valve. Several different designs exist, such as the Geroller (internal or external rollers) and Nichols motors. Typically, the Gerotor motors are low-to-medium speed and medium-to-high torque.

Axial Plunger Motors

For high quality rotating drive systems plunger motors are generally used. Whereas the speed of hydraulic pumps ranges from 1200 to 1800 rpm, the machinery to be driven by the motor often requires a much lower speed. This means that when an axial plunger motor (swept volume maximum 2 litres) is used, a gearbox is usually needed. For a continuously adjustable swept volume, axial piston motors are used. PISTON TYPE.— Like piston (reciprocating) type pumps, the most common design of the piston type of motor is the axial. This type of motor is the most commonly used in hydraulic systems. These motors are, like their pump counterparts, available in both variable and fixed displacement designs. Typical usable (within acceptable efficiency) rotational speeds range from below 50rpm and up to and above 14000rpm. Efficiencies and minimum/maximum rotational speeds are highly dependent on the design of the rotating group, and many different types are in use.

Radial Piston Motors

Radial piston motors are available in two basic types. Crankshaft type (e.g. Staffa or SAI motors) with a single cam and pistons pushing inwards. This type of motor is basically an old design but is one which has extremely high starting torque characteristics. They are available in displacements from 40cc/rev up to about 50 litres/rev but can sometimes be limited in power output. Crankshaft type Radial Piston Motors are capable of running at "creep" speeds and some can run seamlessly up to 1500 rpm whilst offering virtually constant output Torque chacteristics. This makes them still the most versatile design. Multilobe cam ring type (e.g. Hägglunds or Poclain type) has a cam ring with multiple lobes and the piston rollers push outwardly against the cam ring. This produces a very smooth output with high starting torque but they are often limited in the upper speed range. This type of motor is available in a very wide range from about 1 litre/rev to 250 litres/rev. These motors are particularly good on low speed applications and can develop very high power.

Uses

Hydraulic motors are used for many applications now such as winches and crane drives, wheel motors for military vehicles, self-driven cranes, and excavators. Conveyor and feeder drives, mixer and agitator drives, roll mills, drum drives for digesters, trommels and kilns, shredders for cars, tyres, cable and general garbage, drilling rigs, trench cutters, high-powered lawn trimmers, plastic injection machines

Chapter 5

Hydraulic Power Network

A hydraulic power network is a system of interconnected pipes carrying pressurized liquid used to transmit mechanical power from a power source, like a pump, to hydraulic equipment like lifts or motors. The system is analogous to an electrical grid transmitting power from a generating station to end-users. Only a few hydraulic power transmission networks are still in use; modern hydraulic equipment has a pump built into the machine. In the late 19th century, a hydraulic network might have been used in a factory, with a central steam engine or water turbine driving a pump and a system of high-pressure pipes transmitting power to various machines.

The idea of a public hydraulic power network was suggested by Joseph Bramah in a patent obtained in 1812. William Armstrong began installing systems in England from the 1840s, using low-pressure water, but a breakthrough occurred in 1850 with the introduction of the hydraulic accumulator, which allowed much higher pressures to be used. The first public network, supplying many companies, was constructed in Kingston upon Hull, England. The Hull Hydraulic Power Company began operation in 1877, with Edward B. Ellington as its engineer.

Ellington was involved in most of the British networks, and some further afield. Public networks were constructed in Britain at London, Liverpool, Birmingham, Manchester and Glasgow. There were similar networks in Antwerp, Melbourne, Sydney, Buenos Aires and Geneva. All of the public networks had ceased to operate by the mid-1970s, but Bristol Harbour still has an operational system, with an accumulator situated outside the main pumphouse, enabling its operation to be easily visualised.

History

Joseph Bramah, an inventor and locksmith living in London, registered a patent at the London Patent Office on 29 April 1812, which was principally about a provision of a public water supply network, but included a secondary concept for the provision of a high-pressure water main, which would enable workshops to operate machinery. The high-pressure water would be applied "to a variety of other useful purposes, to which the same has never before been so applied." Major components of the system were a ring main, into which a number of pumping stations would pump the water, with pressure being regulated by several air vessels or loaded pistons. Pressure relief valves would protect the system, which he believed could deliver water at a pressure of "a great plurality of atmospheres", and in concept, this was how later hydraulic power systems worked.

In Newcastle upon Tyne, a solicitor called William Armstrong, who had been experimenting with water-powered machines, was working for a firm of solicitors who were appointed to act on behalf of the Whittle Dene Water Company. The water company had been set up to supply Newcastle with drinking water, and Armstrong was appointed secretary at the first meeting of shareholders. Soon afterwards, he wrote to Newcastle Town Council, suggesting that the cranes on the quay should be converted to hydraulic power. He was required to carry out the work at his own expense, but would be rewarded if the conversion was a success. It was, and he set up the Newcastle Cranage Company, which received an order for the conversion of the other four cranes. Further work followed, with the engineer from Liverpool Docks visiting Newcastle and being impressed by a demonstration of the crane's versatility, given by the crane driver John Thorburn, known locally as "Hydraulic Jack".

While the Newcastle system ran on water from the public water supply, the crane installed by Armstrong at Burntisland was not located where such an option was possible, and so he built a 180-foot (55 m) tower, with a water tank at the top, which was filled by a 6 hp (4.5 kW) steam engine. At Elswick in Glasgow, charges by the Corporation Water Department for the water used persuaded the owners that the use of a steam-powered crane would be cheaper. Bramah's concept of "loaded pistons" was introduced in 1850, when the first hydraulic accumulator was installed as part of a scheme for cranes for the Manchester, Sheffield and Lincolnshire Railway. A scheme for cranes at Paddington the following year specified an accumulator with a 10-inch (250 mm) piston and a stroke of 15 feet

(4.6 m), which enabled pressures of 600 pounds per square inch (41 bar) to be achieved. Compared to the 80 psi (5.5 bar) of the Newcastle scheme, this increased pressure significantly reduced the volumes of water used. Cranes were not the only application, and hydraulic operation of the dock gates at Swansea reduced the operating time from 15 to two minutes, and the number of men required to operate them from twelve to four. Each of these schemes was for a single customer, and the application of hydraulic power more generally required a new model.

Public Power

The first practical installation which supplied hydraulic power to the public was in Kingston upon Hull, in England. The Hull Hydraulic Power Company began operation in 1876. They had 2.5 miles (4.0 km) of pipes, which were up to 6 inches (150 mm) in diameter, and ran along the west bank of the River Hull from Sculcoates bridge to its junction with the Humber. The pumping station was near the north end of the pipeline, on Machell Street, near the disused Scott Street bascule bridge, which was powered hydraulically. There was an accumulator at Machell Street, and another one much nearer the Humber, on the corner of Grimsby Lane. Special provision was made where the pressure main passed under the entrance to Queens Dock. By 1895, pumps rated at 250 hp (190 kW) pumped some 500,000 imperial gallons (2,300 m^3) of water into the system each week, and 58 machines were connected to it. The working pressure was 700 psi (48 bar), and the water was used to operate cranes, dock gates, and a variety of other machinery connected with ships and shipbuilding. The Hull system lasted until the 1940s, when the systematic bombing of the city during the Second World War led to the destruction of much of the infrastructure, and the company was wound up in 1947, when Mr F J Haswell, who had been the manager and engineer since 1904, retired.

The man responsible for the Hull system was Edward B. Ellington, who had risen to become the managing director of the Hydraulic Engineering Company, based in Chester, since first joining it in 1869. At the time of its installation, such a scheme seemed like "a leap in the dark", according to R. H. Tweddell writing in 1895, but despite a lack of enthusiasm for the scheme, Ellington pushed ahead and used it as a test bed for both the mechanical and the commercial aspects of the idea. He was eventually involved on some level in most of the hydraulic power networks of Britain. The success of such systems led to them being installed in places as far away as Antwerp in Belgium,

Melbourne and Sydney in Australia, and Buenos Aires in Argentina. The best-known public hydraulic network was the city-wide network of the London Hydraulic Power Company.

This was formed in 1882, as the General Hydraulic Power Company, with Ellington as the consulting engineer. By 1883, another enterprise, the Wharves and Warehouses Steam Power and Hydraulic Pressure Company, had begun to operate, with 7 miles (11 km) of pressure mains on both sides of the River Thames. These supplied cranes, dock gates, and other heavy machinery. Under the terms of an Act of Parliament obtained in 1884, the two companies amalgamated, and became the London Hydraulic Power Company. Initially supplying 17.75 million gallons (80.7 Megalitres) of high-pressure water each day, this had risen to 1,650 million gallons (7,500 Megalitres) by 1927, when the company was powering around 8,000 machines from the supply. They maintained 184 miles (296 km) of mains at 700 psi (48 bar), which covered an area reaching Pentonville in the north, Limehouse in the east, Nine Elms and Bermondsey in the south and Earls Court and Notting Hill in the west.

Five pumping stations kept the mains pressurised, assisted by accumulators. The original station was at Falcon Wharf, Bankside, but this was replaced by four stations at Wapping, Rotherhithe, Grosvenor Road in Pimlico and City Road in Clerkenwell. A fifth station at East India Docks was originally operated by the Port of London Authority, but was taken over and connected to the system. The stations used steam engines until 1953, when Grosvenor Road station was converted to use electric motors, and following the success of this project, the other four were also converted.

The electric motors allowed much smaller accumulators to be used, since they were then only controlling the pressure and flow, rather than storing power. While the network supplied lifts, cranes and dockgates, it also powered the cabaret platform at the Savoy Hotel, and from 1937, the 720-tonne three-section central floor at the Earls Court Exhibition Centre, which could be raised or lowered relative to the main floor to convert between a swimming pool and an exhibition hall. The London system contracted during the Second World War, due to the destruction of customer's machinery and premises. Following the hostilities, large areas of London were reconstructed, and the re-routing of pressure mains was much more difficult than the provision of an electric supply, so that by 1954 the number of machines had fallen to 4286. The company was wound up in 1977.

A system began operating in Liverpool in 1888. It was an offshoot of the London-based General Hydraulic Power Company, and was authorised by acts of Parliament obtained in 1884 and 1887. By 1890, some 16 miles (26 km) of mains had been installed, supplied by a pumping station at Athol Street, on the bank of the Leeds and Liverpool Canal. Although water was originally taken from the canal, cleaner water supplied by Liverpool Corporation was in use by 1890, removing the need for a filtration plant. At this time two pumpsets were in use, and a third was being installed. Pressure was maintained by two accumulators, each with an 18-inch (460 mm) diameter piston with a stroke of 20 feet (6.1 m). *The Practical Engineer* quoted the pressure as 75 pounds per square inch (5.2 bar), but this is unlikely to be correct by comparison with other systems. A second pumping station at Grafton Street was operational by 1909. The system ceased operation in 1971.

Birmingham obtained its system in 1891, when the Dalton Street hydraulic station opened. In an unusual move, J. W. Gray, the Water Department engineer for the city, had been laying pressure mains beneath the streets for some years, anticipating the need for such a system. The hydraulic station used Otto 'Silent' type gas engines, and had two accumulators, with an 18-inch (460 mm) diameter piston, a stroke of 20 feet (6.1 m) and each loaded with a 93-tonne weight. The gas engines were started by a small hydraulic engine, which used the hydraulic energy stored in the accumulators, and all equipment was supplied by Ellington's company. Very few documents describing the details of the system are known to exist.

The final two public systems in Britain were in Manchester, commissioned in 1894, and Glasgow, commissioned the following year. Both were equipped by Ellington's company, and used the higher pressure of 1,120 psi (77 bar). This was maintained by six sets of triple-expansion steam engines, rated at 200 hp (150 kW) each. Two accumulators with pistons of 18-inch (460 mm) diameter, a stroke of 23 feet (7.0 m), and loaded with 127 tonnes were installed. In Manchester, the hydraulic station was built on the east side of Gloucester Street, by Oxford Road railway station. It was later supplemented by stations at Water Street and Pott Street, the latter now under the car parks of the Central Retail Park. At its peak in the 1930s, the system consisted on 35 miles (56 km) of pipes, which were connected to 2,400 machines, most of which were used for baling cotton. The system was shut down in 1972. In Glasgow, the pumping station was at the junction of High Street and Rottenrow. By 1899, it was supplying power to 348 machines, and another 39 were in the process of being completed. The pipes were 7 inches (180 mm) in diameter, and there were around 30 miles (48 km) of them by 1909, when 202,141 imperial gallons (918.95 m^3) of high pressure water were supplied to customers. The system was shut down in 1964.

Other Systems

All of the British systems were designed to provide power for intermittent processes, such as the operation of dock gates or cranes. The system installed at Antwerp was somewhat different, in that its primary purpose was the production of electricity for lighting. It was commissioned in 1894, and used pumping engines producing a total of 1,000 hp (750 kW) to supply water at 750 psi (52 bar). Ellington, writing in 1895, stated that he found it difficult to see that this was an economical use of hydraulic power, although tests conducted at his works at Chester in October 1894 showed that efficiencies of 59 per cent could be achieved using a Pelton wheel directly coupled to a dynamo.

Two major systems were built in Australia. The first was in Melbourne, where the Melbourne Hydraulic Power Company began operating in July 1889. The company was authorised by an Act of the Victorian Parliament passed in December 1887, and construction of the system began, with Coates & Co. acting as consulting engineers, and George Swinburne working as engineering manager. The steam pumping plant was supplied by Abbot & Co. from England. Expansion was rapid, with around 70 machines, mainly hydraulic lifts, connected to the system by the end of 1889, and a third steam engine had to be installed in mid-1890, which more than doubled the capacity of the system. A fourth pumping engine was added in 1891, by which time there were 100 customers connected to the mains. The mains were a mixture of 4-inch (100 mm) and 6-inch (150 mm) pipes. The water was extracted from the Yarra River until 1893, after which it was drawn from the Public Works Department's supply. There were some 16 miles (26 km) of mains by 1897. A second pumping station was added in 1901, and in 1902, 102 million gallons (454 Megalitres) of pressurised water were used by customers.

The system was operated as a commercial enterprise until 1925, after which the business and its assets reverted to the City of Melbourne, as specified by the original act. One of the early improvements made by the City Council was to consolidate the system. The steam pumps were replaced by new electric pumps, located in the

Spencer Street power station, which thus supplied both electric power and hydraulic power to the city. The hydraulic system continued to operate under municipal ownership until December 1967.

In January 1891, a system in Sydney came on-line, having been authorised by act of Parliament in 1888. George Swinburne was again the engineer, and the system was supplying power to around 200 machines by 1894, which included 149 lifts and 20 dock cranes. The operating company was the Sydney and Suburbs Hydraulic Power Company, later shortened to the Sydney Hydraulic Power Company. Pressure mains were either of 4-inch (100 mm) or 6-inch (150 mm) diameter, and at its peak, there were around 50 miles (80 km) of mains, covering an area between Pyrmont, Woolloomooloo and Broadway. In 1919, most of the 2369 lifts in the metropolitan area were hydraulically operated. The pumping station, together with two accumulators, was situated in the Darling Harbour district, and the original steam engines were replaced by three electric motors driving centrifugal pumps in 1952. The scheme remained in private ownership until its demise in 1975, and the pumping station has since been re-used as a tavern.

Ellington's system in Buenos Aires was designed to operate a sewage pumping scheme in the city.

Summary

System	*Operational*	*Closed*	*Pumping Stations*	*Mains (miles)*	*Mains (km)*	*Pressure (psi)*	*Pressure (bar)*
Hull	1876	1947	1	2.5	4	700	48
London	1883	1977	5	184	296	750	52
Liverpool	1888	1971	2	30	48	800	55
Melbourne	1889	1967	2	16	26	750	52
Birmingham	1891		1			700	48
Sydney	1891	1975	1	50	80	750	52
Manchester	1894	1972	3	35	56	1,120	77
Antwerp	1894		1	4.5	7.2	750	52
Glasgow	1895	1964	1	30	48	1,120	77

Legacy

Geneva had a hydraulic network with a total installed power of 6,000 hp (4,500 kW). The system gave birth to the Jet d'Eau, a huge fountain which was originally only a pressure relief valve for the network when the consumers had been switched off.

Bristol Harbour still has a working system, the pumping machinery of which was supplied by Fullerton, Hodgart and Barclay of Paisley, Scotland in 1907. The engine house is a grade II* listed building, constructed in 1888, with a tower at one end to house the hydraulic accumulator. A second accumulator was fitted outside the building in the 1920s, which enables the operation of the system to be more easily visualised. A number of artifacts, including the buildings used as pumping stations, have survived the demise of public hydraulic power networks. In Hull, the Machell Street pumping station has been reused as a workshop. The building still supports the sectional cast-iron roof tank used to allow the silt-laden water of the River Hull to settle, and is marked by a Blue plaque, to commemorate its importance. In London, Bermondsey pumping station, built in 1902, is in use as an engineering works, but retains its chimney and accumulator tower, while the station at Wapping is virtually complete, retaining all of its equipment, which is still in working order. The building is grade II* listed because of its completeness.

In Manchester, the Water Street pumping station, built in Baroque style between 1907 and 1909, was used as workshops for the City College, but has formed part of the People's History Museum since 1994. One of the pumping sets has been moved to the Museum of Science and Industry, where it has been restored to working order and forms part of a larger display about hydraulic power. The pumps were made by the Manchester firm of Galloways.

Hydraulic Press

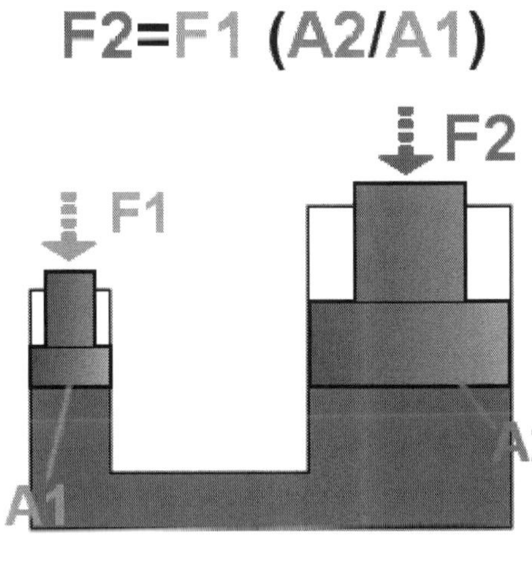

Figure: *Hydraulic force increase*

A hydraulic press is a machine using a hydraulic cylinder to generate a compressive force. It uses the hydraulic equivalent of a mechanical lever, and was also known as a Bramah press after the inventor, Joseph Bramah, of England. He invented and was issued a patent on this press in 1795. As Bramah (who is also known for his development of the flush toilet) installed toilets, he studied the existing literature on the motion of fluids and put this knowledge into the development of the press.

Principle

The hydraulic press depends on Pascal's principle: the pressure throughout a closed system is constant. One part of the system is a piston acting as a pump, with a modest mechanical force acting on a small cross-sectional area; the other part is a piston with a larger area which generates a correspondingly large mechanical force. Only small-diameter tubing (which more easily resists pressure) is needed if the pump is separated from the press cylinder.

Pascal's law: Pressure on a confined fluid is transmitted undiminished and acts with equal force on equal areas and at 90 degrees to the container wall.

A fluid, such as oil, is displaced when either piston is pushed inward. The small piston, for a given distance of movement, displaces a smaller amount of volume than the large piston, which is proportional to the ratio of areas of the heads of the pistons. Therefore, the small piston must be moved a large distance to get the large piston to move significantly. The distance the large piston will move is the distance that the small piston is moved divided by the ratio of the areas of the heads of the pistons. This is how energy, in the form of work in this case, is conserved and the Law of Conservation of Energy is satisfied. Work is force times distance, and since the force is increased on the larger piston, the distance the force is applied over must be decreased.

Applications

Hydraulic presses are commonly used for forging, moulding, blanking, punching, deep drawing, and metal forming operations.

In Popular Culture

A hydraulic press is the cause of death for the Terminator in the film of the same name, as well as Andre Delambre in *The Fly*. The room featured in *Fermat's Room* has a design similar to that of a hydraulic press. Boris Artzybasheff also created a drawing of a hydraulic press, in which the press was created out of the shape of a robot.

Hydraulic Rescue Tools

Hydraulic rescue tools are used by emergency rescue personnel to assist vehicle extrication of crash victims, as well as other rescues from small spaces. These tools include cutters, spreaders, door busters and rams. They are popularly referred to in the English speaking Commonwealth nations and the US as the Jaws of Life, a trademark of Hurst Performance, Inc. The Jaws of Life was first used in 1963 as a tool to free race car drivers from their vehicles after accidents.

Hydraulic rescue tools are powered by a hydraulic pump, which can be hand-, foot-, or engine-powered, or even built into the tool. These tools may be either single-acting, where hydraulic pressure will only move the cylinder in one direction, and the return to starting position is accomplished using a pressure-relief valve and spring setup, or dual-acting, in which hydraulic pressure is used to both open and close the hydraulic cylinder.

History

Previously rescuers often used circular saws for vehicle extrication, but these suffered from several drawbacks. Saws can conduct sparks, which could start a fire, create loud sounds, stress the victim(s), and often cut slowly. Alternatively, rescuers could try to pry open the vehicle doors with a crowbar or Halligan bar, but this could compromise the stability of the vehicle, injure the victims further, or inadvertently trigger the airbags of the vehicle.

In comparison, hydraulic spreader-cutters are quieter, faster, stronger, and more versatile: they can cut, open, and even lift a vehicle. *Jaws of Life* is a now trademarked line of tools originally developed by Hurst Performance. Hurst Performance began to export parts to a European company to avoid import duty. In partnership with Zumro resQtec for use in auto racing, with Hurst targeting the American market and resQtec targeting the European market. The hydraulic spreader was originally developed in 1972 by Tim Smith and Mike Brick, who later developed a cutter and a hydraulic ram. When an occupant is trapped the tool is used to pry or cut the car to remove the occupant. It takes about two minutes to take the roof off a car. Mike Brick coined the phrase "Jaws of Life" after he observed people saying that their new device "snatched people from the jaws of death." Later Brick went on to develop a single rescue tool that could perform all the functions of rescue — push, pull, cut and spread. His patent for this unique design eventually led to the Phoenix Rescue tool, which incorporates his design to this day. Other manufacturers (including Holmatro and Lukas, amongst others) make equivalent tools.

Tool Types: Cutter

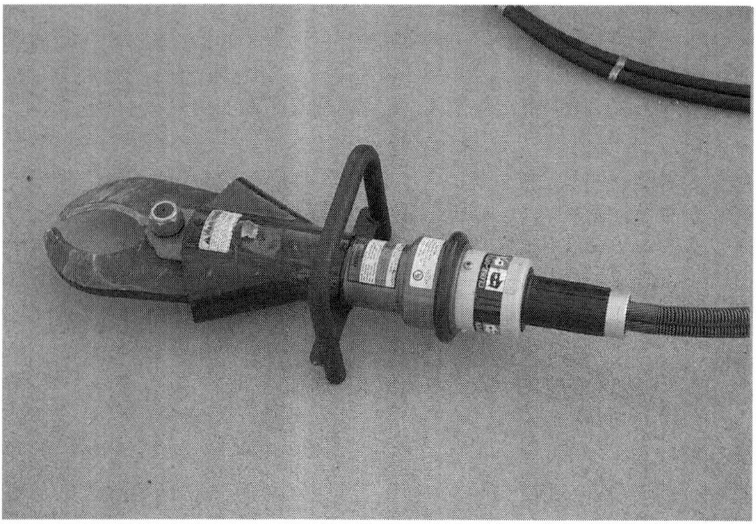

***Figure:** Hydraulic shears*

The cutter is a hydraulic tool which is designed to cut through metal—a pair of hydraulically powered shears. It is often called the jaws of life, owing to the shape and configuration of its blades. Sometimes specified as to its capacity to cut a solid circular steel bar, these are most commonly used to cut through a vehicle's structure in an extraction operation. Cutter blades are replaceable, and blade development progresses as vehicle technology progresses in order to be able to cope with the new car protection technology.

Spreader

A spreader is a hydraulic tool designed with two arms which come together in a narrow tip, and which uses hydraulic pressure to separate or spread the arms. The tip of the tool can be inserted into a narrow gap between two vehicle panels (such as between two doors, or between a door and a fender) — when the tool is operated, the arms are opened, pushing apart the metal in the panels. Spreaders may also be used to "pop" vehicle doors from their hinges.

Spreader-Cutters

Previously rescuers often used circular saws for vehicle extrication, but these suffered from several drawbacks. Saws can conduct sparks, which could start a fire, create loud sounds, stress the victim(s), and often cut slowly. Alternatively, rescuers could try to pry open the vehicle doors with a crowbar or Halligan bar, but this could compromise the stability of the vehicle, injure the victims further, or inadvertently trigger the airbags of the vehicle.

In comparison, hydraulic spreader-cutters are quieter, faster, stronger, and more versatile: they can cut, open, and even lift a vehicle. *Jaws of Life* is a now trademarked line of tools originally developed by Hurst Performance. Hurst Performance began to export parts to a European company to avoid import duty. In partnership with Zumro resQtec for use in auto racing, with Hurst targeting the American market and resQtec targeting the European market. The hydraulic spreader was originally developed in 1972 by Tim Smith and Mike Brick, who later developed a cutter and a hydraulic ram. When an occupant is trapped the tool is used to pry or cut the car to remove the occupant. It takes about two minutes to take the roof off a car. Mike Brick coined the phrase "Jaws of Life" after he observed people saying that their new device "snatched people from the jaws of death." Later Brick went on to develop a single rescue tool that could perform all the functions of rescue — push, pull, cut and spread. His patent for this unique design eventually led to the Phoenix Rescue tool, which incorporates his design to this day. Other manufacturers (including Holmatro and Lukas, amongst others) make equivalent tools.

Rams

***Figure:** An Amkus Ram*

Rams are used far less than spreader-cutters in auto rescues; nonetheless, they serve an important purpose. There are many types and sizes, including single-piston, dual-piston and telescopic rams.

Sizes commonly vary from 50.80 cm (20") to 178 cm (70") (extended). Rams use more hydraulic fluid during operation than spreader-cutters, so it is essential that the pump being used has enough capacity to allow the ram to reach full extension.

Power

The tools operate on the basis of hydraulic fluid pressure of up to 720 bar (10,000 psi), which must be provided from a power source. At present, there are three different means of generating the pressure. The most commonly used source is a separate power unit, a small petrol (gasoline) engine connected to a hydraulic pump. The hydraulic fluid is pressurised in the pump, and conveyed in a hose under pressure to the tool. Alternative power sources are a small electrical pump in the tool powered by a heavy duty rechargeable battery, and a hand- or foot-operated pump also connected to the tool directly. These are useful for lighter-duty use and do not require the storage space taken up by the separate power unit and associated lengths of hose.

Wave

In physics a wave is a disturbance or oscillation that travels through space and matter, accompanied by a transfer of energy. Wave motion transfers energy from one point to another, often with no permanent displacement of the particles of the medium—that is, with little or no associated mass transport. They consist, instead, of oscillations or vibrations around almost fixed locations. Waves are described by a wave equation which sets out how the disturbance proceeds over time. The mathematical form of this equation varies depending on the type of wave. There are two main types of waves. Mechanical waves propagate through a medium, and the substance of this medium is deformed. The deformation reverses itself owing to restoring forces resulting from its deformation. For example, sound waves propagate via air molecules colliding with their neighbours. When air molecules collide, they also bounce away from each other (a restoring force). This keeps the molecules from continuing to travel in the direction of the wave.

The second main type of wave, electromagnetic waves, do not require a medium. Instead, they consist of periodic oscillations of electrical and magnetic fields generated by charged particles, and can therefore travel through a vacuum. These types of waves vary in wavelength, and include radio waves, microwaves, infrared radiation, visible light, ultraviolet radiation, X-rays, and gamma rays.

Further, the behaviour of particles in quantum mechanics is described by waves and researchers believe that gravitational waves also travel through space, although gravitational waves have never been directly detected. A wave can be transverse or longitudinal depending on the direction of its oscillation. Transverse waves occur when a disturbance creates oscillations perpendicular (at right angles) to the propagation (the direction of energy transfer). Longitudinal waves occur when the oscillations are parallel to the direction of propagation. While mechanical waves can be both transverse and longitudinal, all electromagnetic waves are transverse.

General Features

A single, all-encompassing definition for the term *wave* is not straightforward. A vibration can be defined as a *back-and-forth* motion around a reference value. However, a vibration is not necessarily a wave. An attempt to define the necessary and sufficient characteristics that qualify a phenomenon to be called a *wave* results in a fuzzy border line.

The term *wave* is often intuitively understood as referring to a transport of spatial disturbances that are generally not accompanied by a motion of the medium occupying this space as a whole. In a wave, the energy of a vibration is moving away from the source in the form of a disturbance within the surrounding medium (Hall 1980, p. 8). However, this notion is problematic for a standing wave (for example, a wave on a string), where energy is moving in both directions equally, or for electromagnetic (e.g., light) waves in a vacuum, where the concept of medium does not apply and interaction with a target is the key to wave detection and practical applications. There are water waves on the ocean surface; gamma waves and light waves emitted by the Sun; microwaves used in microwave ovens and in radar equipment; radio waves broadcast by radio stations; and sound waves generated by radio receivers, telephone handsets and living creatures (as voices), to mention only a few wave phenomena.

It may appear that the description of waves is closely related to their physical origin for each specific instance of a wave process. For example, acoustics is distinguished from optics in that sound waves are related to a mechanical rather than an electromagnetic wave transfer caused by vibration. Concepts such as mass, momentum, inertia, or elasticity, become therefore crucial in describing acoustic (as distinct from optic) wave processes. This difference in origin introduces certain wave characteristics particular to the properties of the medium involved. For example, in the case of air: vortices, radiation pressure, shock waves etc.; in the case of solids: Rayleigh waves, dispersion; and so on.

Other properties, however, although usually described in terms of origin, may be generalized to all waves. For such reasons, wave theory represents a particular branch of physics that is concerned with the properties of wave processes independently of their physical origin. For example, based on the mechanical origin of acoustic waves, a moving disturbance in space–time can exist if and only if the medium involved is neither infinitely stiff nor infinitely pliable. If all the parts making up a medium were rigidly *bound*, then they would all vibrate as one, with no delay in the transmission of the vibration and therefore no wave motion. On the other hand, if all the parts were independent, then there would not be any transmission of the vibration and again, no wave motion. Although the above statements are meaningless in the case of waves that do not require a medium, they reveal a characteristic that is relevant to all waves regardless of origin: within a wave, the phase of a vibration (that is, its position within the vibration cycle) is different for adjacent points in space because the vibration reaches these points at different times.

Similarly, wave processes revealed from the study of waves other than sound waves can be significant to the understanding of sound phenomena. A relevant example is Thomas Young's principle of interference (Young, 1802, in Hunt 1992, p. 132). This principle was first introduced in Young's study of light and, within some specific contexts (for example, scattering of sound by sound), is still a researched area in the study of sound.

Mathematical Description of One-Dimensional Waves

Wave Equation

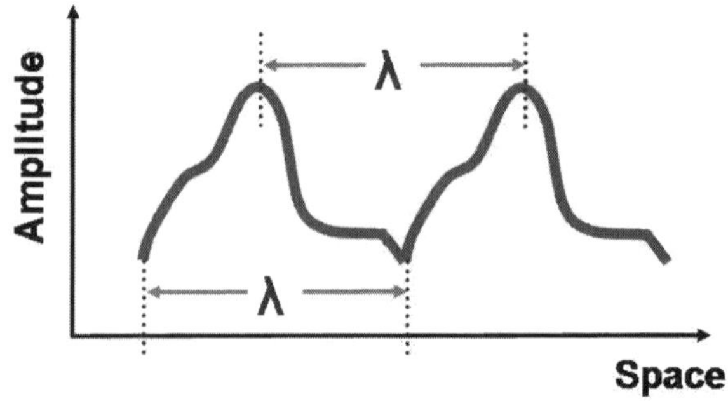

Figure: *Wavelength λ, can be measured between any two corresponding points on a waveform*

Consider a travelling transverse wave (which may be a pulse) on a string (the medium). Consider the string to have a single spatial dimension. Consider this wave as travelling

- in the x direction in space. E.g., let the positive direction be to the right, and the negative x direction be to the left.
- with constant amplitude u
- with constant velocity u, where u is
  - o independent of wavelength (no dispersion)
  - o independent of amplitude (linear media, not nonlinear).
- with constant waveform, or shape

This wave can then be described by the two-dimensional functions

$u(x,t) = F(x - v\,t)$ (waveform F travelling to the right)

$u(x,t) = G(x + v\,t)$ (waveform G travelling to the left)

or, more generally, by d'Alembert's formula:

$$u(x,t) = F(x - vt) + G(x + vt).$$

representing two component waveforms F and G travelling through the medium in opposite directions. A generalized representation of this wave can be obtained as the partial differential equation

$$\frac{1}{v^2}\frac{\partial^2 u}{\partial t^2} = \frac{\partial^2 u}{\partial x^2}.$$

General solutions are based upon Duhamel's principle.

Wave Forms

The form or shape of F in d'Alembert's formula involves the argument $x - vt$. Constant values of this argument correspond to constant values of F, and these constant values occur if x increases at the same rate that vt increases. That is, the wave shaped like the function F will move in the positive x-direction at velocity v (and G will propagate at the same speed in the negative x-direction).

In the case of a periodic function F with period λ, that is, $F(x + \lambda - vt) = F(x - vt)$, the periodicity of F in space means that a snapshot of the wave at a given time t finds the wave varying periodically in space with period λ (the wavelength of the wave). In a similar fashion, this periodicity of F implies a periodicity in time as well: $F(x - v(t + T)) = F(x - vt)$ provided $vT = \lambda$, so an observation of the wave at a fixed location x finds the wave undulating periodically in time with period $T = \lambda/v$.

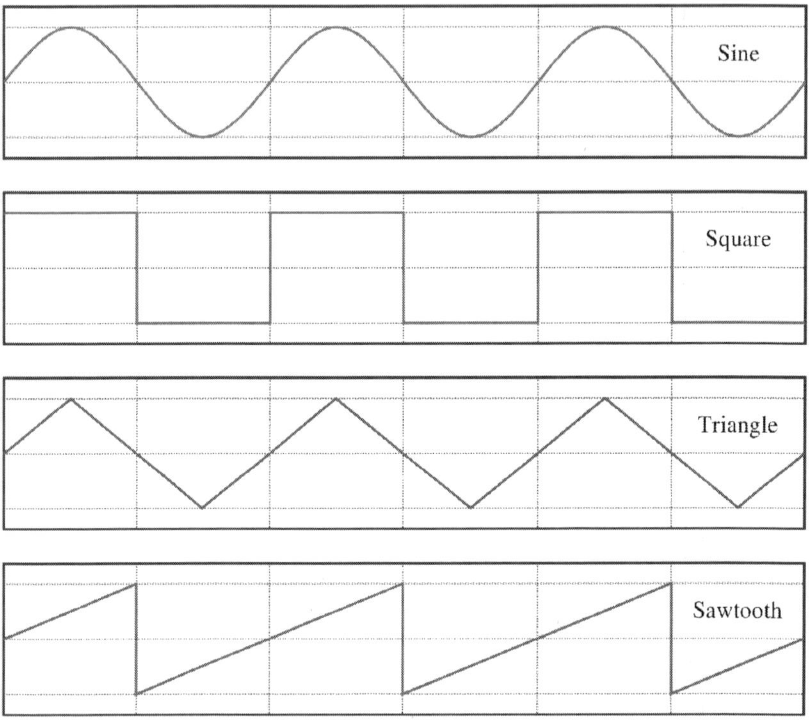

Figure: *Sine, square, triangle and sawtooth waveforms.*

Amplitude and Modulation

The amplitude of a wave may be constant (in which case the wave is a *c.w.* or *continuous wave*), or may be *modulated* so as to vary with time and/or position.

The outline of the variation in amplitude is called the *envelope* of the wave. Mathematically, the modulated wave can be written in the form:

$$u(x,t) = A(x,t)\sin(kx - \omega t + \phi),$$

where $A(x,t)$ is the amplitude envelope of the wave, k is the *wavenumber* and is the *phase*. If the group velocity v_g is wavelength-independent, this equation can be simplified as:

$$u(x,t) = A(x - v_g\ t)\sin(kx - \omega t + \phi),$$

showing that the envelope moves with the group velocity and retains its shape. Otherwise, in cases where the group velocity varies with wavelength, the pulse shape changes in a manner often described using an *envelope equation.*

Phase Velocity and Group Velocity

There are two velocities that are associated with waves, the phase velocity and the group velocity. To understand them, one must consider several types of waveform. For simplification, examination is restricted to one dimension.

The most basic wave (a form of plane wave) may be expressed in the form:

$$\psi(x,t) = Ae^{i(kx-\omega t)},$$

which can be related to the usual sine and cosine forms using Euler's formula. Rewriting the argument, $kx - \omega t = \left(\frac{2\pi}{\lambda}\right)(x - vt)$, makes clear that this expression describes a vibration of wavelength $\lambda = \frac{2\pi}{k}$ travelling in the *x*-direction with a constant *phase velocity* $v_p = \frac{\omega}{k}$.

The other type of wave to be considered is one with localized structure described by an envelope, which may be expressed mathematically as, for example:

$$\psi(x,t) = \int_{-\infty}^{\infty} dk_1\ A(k_1)\, e^{i(k_1 x-\omega t)},$$

where now $A(k_1)$ (the integral is the inverse fourier transform of A(k1)) is a function exhibiting a sharp peak in a region of wave vectors Δk surrounding the point $k_1 = k$. In exponential form:

$$A = A_o(k_1)e^{i\alpha(k_1)},$$

with A_0 the magnitude of A. For example, a common choice for A_0 is a Gaussian wave packet:

$$A_o(k_1) = N\, e^{-\sigma^2(k_1-k)^2/2},$$

where σ determines the spread of k_1-values about k, and N is the amplitude of the wave.

The exponential function inside the integral for ψ oscillates rapidly with its argument, say $\varphi(k_1)$, and where it varies rapidly, the exponentials cancel each other out, interfere destructively, contributing little to ψ. However, an exception occurs at the location where the argument φ of the exponential varies slowly. (This observation is the basis for the method of stationary phase for evaluation of such integrals.) The condition for φ to vary slowly is that its rate of change with k_1 be small; this rate of variation is:

$$\left.\frac{d\varphi}{dk_1}\right|_{k_1=k} = x - t\left.\frac{d\omega}{dk_1}\right|_{k_1=k} + \left.\frac{d\alpha}{dk_1}\right|_{k_1=k},$$

where the evaluation is made at $k_1 = k$ because $A(k_1)$ is centred there. This result shows that the position x where the phase changes slowly, the position where ψ is appreciable, moves with time at a speed called the *group velocity*:

$$v_g = \frac{d\omega}{dk}.$$

The group velocity therefore depends upon the dispersion relation connecting ω and k. For example, in quantum mechanics the energy of a particle represented as a wave packet is $E = \hbar\omega = (\hbar k)^2/(2m)$. Consequently, for that wave situation, the group velocity is

$$v_g = \frac{\hbar k}{m},$$

showing that the velocity of a localized particle in quantum mechanics is its group velocity. Because the group velocity varies with k, the shape of the wave packet broadens with time, and the particle becomes less localized. In other words, the velocity of the constituent waves of the wave packet travel at a rate that varies with their wavelength, so some move faster than others, and they cannot maintain the same interference pattern as the wave propagates.

Sinusoidal Waves

Mathematically, the most basic wave is the (spatially) one-dimensional sine wave (or *harmonic wave* or *sinusoid*) with an amplitude u described by the equation:

$$u(x,t) = A\sin(kx - \omega t + \phi),$$

where

- A is the maximum amplitude of the wave, maximum distance from the highest point of the disturbance in the medium (the crest) to the equilibrium point during one wave cycle. In the illustration to the right, this is the maximum vertical distance between the baseline and the wave.
- x is the space coordinate
- t is the time coordinate
- k is the wavenumber
- ω is the angular frequency
- ϕ is the phase constant.

The units of the amplitude depend on the type of wave. Transverse mechanical waves (e.g., a wave on a string) have an amplitude expressed as a distance (e.g., metres), longitudinal mechanical waves (e.g., sound waves) use units of pressure (e.g., pascals), and electromagnetic waves (a form of transverse vacuum wave) express the amplitude in terms of its electric field (e.g., volts/metre).

The wavelength λis the distance between two sequential crests or troughs (or other equivalent points), generally is measured in metres. A wavenumber *k*, the spatial frequency of the wave in radians per unit distance (typically per metre), can be associated with the wavelength by the relation

$$k = \frac{2\pi}{\lambda}.$$

The *T* period is the time for one complete cycle of an oscillation of a wave. The frequency *f* is the number of periods per unit time (per second) and is typically measured in hertz. These are related by:

$$f = \frac{1}{T}.$$

In other words, the frequency and period of a wave are reciprocals.

The angular frequency ω represents the frequency in radians per second. It is related to the frequency or period by

$$\omega = 2\pi f = \frac{2\pi}{T}.$$

The wavelength λ of a sinusoidal waveform travelling at constant speed v is given by:

$$\lambda = \frac{v}{f},$$

where is v called the phase speed (magnitude of the phase velocity) of the wave and *f* is the wave's frequency.

Wavelength can be a useful concept even if the wave is not periodic in space. For example, in an ocean wave approaching shore, the incoming wave undulates with a varying *local* wavelength that depends in part on the depth of the sea floor compared to the wave height. The analysis of the wave can be based upon comparison of the local wavelength with the local water depth.

Although arbitrary wave shapes will propagate unchanged in lossless linear time-invariant systems, in the presence of dispersion the sine wave is the unique shape that will propagate unchanged but for phase and amplitude, making it easy to analyze. Due to the Kramers–Kronig relations, a linear medium with dispersion also exhibits loss, so the sine wave propagating in a dispersive medium is attenuated in certain frequency ranges that depend upon the medium. The sine function is periodic, so the sine wave or sinusoid has a wavelength in space and a period in time.

The sinusoid is defined for all times and distances, whereas in physical situations we usually deal with waves that exist for a limited span in space and duration in time. Fortunately, an arbitrary wave shape can be decomposed into an infinite set of sinusoidal waves by the use of Fourier analysis.

As a result, the simple case of a single sinusoidal wave can be applied to more general cases. In particular, many media are linear, or nearly, so the calculation of arbitrary wave behaviour can be found by adding up responses to individual sinusoidal waves using the superposition principle to find the solution for a general waveform. When a medium is nonlinear, the response to complex waves cannot be determined from a sine-wave decomposition.

Plane Waves

Standing Waves: A standing wave, also known as a *stationary wave*, is a wave that remains in a constant position. This phenomenon can occur because the medium is moving in the opposite direction to the wave, or it can arise in a stationary medium as a result of interference between two waves travelling in opposite directions.

The *sum* of two counter-propagating waves (of equal amplitude and frequency) creates a *standing wave*. Standing waves commonly arise when a boundary blocks further propagation of the wave, thus causing wave reflection, and therefore introducing a counter-propagating wave.

For example when a violin string is displaced, transverse waves propagate out to where the string is held in place at the bridge and the nut, where the waves are reflected back. At the bridge and nut, the two opposed waves are in antiphase and cancel each other, producing a node. Halfway between two nodes there is an antinode, where the two counter-propagating waves *enhance* each other maximally. There is no net propagation of energy over time.

Physical Properties

Waves exhibit common behaviours under a number of standard situations, e. g.

Transmission and Media

Waves normally move in a straight line (i.e. rectilinearly) through a *transmission medium*. Such media can be classified into one or more of the following categories:

- A *bounded medium* if it is finite in extent, otherwise an *unbounded medium*
- A *linear medium* if the amplitudes of different waves at any particular point in the medium can be added
- A *uniform medium* or *homogeneous medium* if its physical properties are unchanged at different locations in space
- An *anisotropic medium* if one or more of its physical properties differ in one or more directions
- An *isotropic medium* if its physical properties are the *same* in all directions

Absorption

Absorption of waves mean, if a kind of wave strikes a matter, it will be absorbed by the matter. When a wave with that same natural frequency impinges upon an atom, then the electrons of that atom will be set into vibrational motion. If a wave of a given frequency strikes a material with electrons having the same vibrational frequencies, then those electrons will absorb the energy of the wave and transform it into vibrational motion.

Reflection

When a wave strikes a reflective surface, it changes direction, such that the angle made by the incident wave and line normal to the surface equals the angle made by the reflected wave and the same normal line.

Interference

Waves that encounter each other combine through superposition to create a new wave called an interference pattern. Important interference patterns occur for waves that are in phase.

Refraction

Refraction is the phenomenon of a wave changing its speed. Mathematically, this means that the size of the phase velocity changes. Typically, refraction occurs when a wave passes from one medium into another. The amount by which a wave is refracted by a material is given by the refractive index of the material. The directions of incidence and refraction are related to the refractive indices of the two materials by Snell's law.

Diffraction

A wave exhibits diffraction when it encounters an obstacle that bends the wave or when it spreads after emerging from an opening. Diffraction effects are more pronounced when the size of the obstacle or opening is comparable to the wavelength of the wave.

Polarization

A wave is polarized if it oscillates in one direction or plane. A wave can be polarized by the use of a polarizing filter. The polarization of a transverse wave describes the direction of oscillation in the plane perpendicular to the direction of travel.

Longitudinal waves such as sound waves do not exhibit polarization. For these waves the direction of oscillation is along the direction of travel.

Dispersion

A wave undergoes dispersion when either the phase velocity or the group velocity depends on the wave frequency. Dispersion is most easily seen by letting white light pass through a prism, the result of which is to produce the spectrum of colours of the rainbow. Isaac Newton performed experiments with light and prisms, presenting his findings in the *Opticks* (1704) that white light consists of several colours and that these colours cannot be decomposed any further.

Mechanical Waves

Waves on strings: The speed of a transverse wave travelling along a vibrating string (v) is directly proportional to the square root of the tension of the string (T) over the linear mass density (μ):

$$v = \sqrt{\frac{T}{\mu}},$$

where the linear density μ is the mass per unit length of the string.

Acoustic Waves

Acoustic or sound waves travel at speed given by

$$v = \sqrt{\frac{B}{\rho_0}},$$

or the square root of the adiabatic bulk modulus divided by the ambient fluid density.

Water Waves

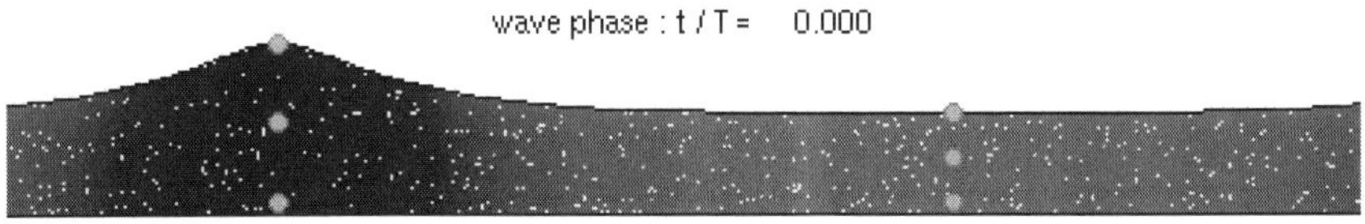

- Ripples on the surface of a pond are actually a combination of transverse and longitudinal waves; therefore, the points on the surface follow orbital paths.
- Sound—a mechanical wave that propagates through gases, liquids, solids and plasmas;
- Inertial waves, which occur in rotating fluids and are restored by the Coriolis effect;
- Ocean surface waves, which are perturbations that propagate through water.

Seismic Waves : Shock Waves

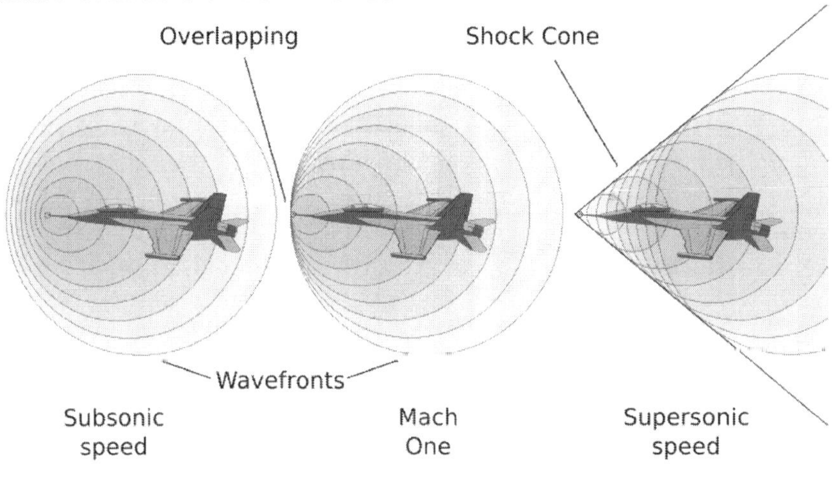

Other

- Waves of traffic, that is, propagation of different densities of motor vehicles, and so forth, which can be modelled as kinematic waves
- Metachronal wave refers to the appearance of a travelling wave produced by coordinated sequential actions.
- It is worth noting that the mass-energy equivalence equation can be solved for this form: $c = \sqrt{\frac{e}{m}}$.

Electromagnetic Waves

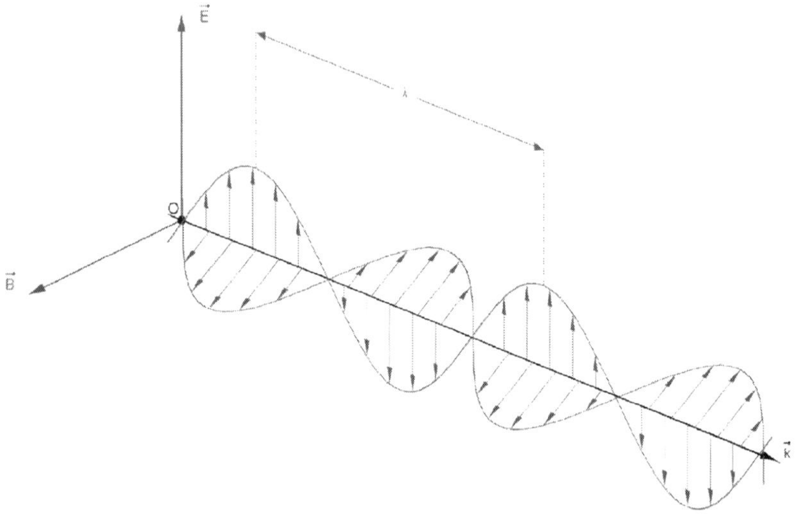

An electromagnetic wave consists of two waves that are oscillations of the electric and magnetic fields. An electromagnetic wave travels in a direction that is at right angles to the oscillation direction of both fields. In the 19th century, James Clerk Maxwell showed that, in vacuum, the electric and magnetic fields satisfy the wave equation both with speed equal to that of the speed of light. From this emerged the idea that light is an electromagnetic wave. Electromagnetic waves can have different frequencies (and thus wavelengths), giving rise to various types of radiation such as radio waves, microwaves, infrared, visible light, ultraviolet and X-rays.

Quantum Mechanical Waves

The Schrödinger equation describes the wave-like behaviour of particles in quantum mechanics. Solutions of this equation are wave functions which can be used to describe the probability density of a particle.

De Broglie Waves

Louis de Broglie postulated that all particles with momentum have a wavelength

$$\lambda = \frac{h}{p},$$

where h is Planck's constant, and p is the magnitude of the momentum of the particle. This hypothesis was at the basis of quantum mechanics.

Nowadays, this wavelength is called the de Broglie wavelength. For example, the electrons in a CRT display have a de Broglie wavelength of about 10^{-13} m.

A wave representing such a particle travelling in the *k*-direction is expressed by the wave function as follows:

$$\psi(\mathrm{r}, t=0) = A\, e^{i\mathrm{k}\cdot\mathrm{r}},$$

where the wavelength is determined by the wave vector k as:

$$\lambda = \frac{2\pi}{k},$$

and the momentum by:

$$\mathrm{p} = \hbar \mathrm{k}.$$

However, a wave like this with definite wavelength is not localized in space, and so cannot represent a particle localized in space. To localize a particle, de Broglie proposed a superposition of different wavelengths ranging around a central value in a wave packet, a waveform often used in quantum mechanics to describe the wave function of a particle. In a wave packet, the wavelength of the particle is not precise, and the local wavelength deviates on either side of the main wavelength value.

In representing the wave function of a localized particle, the wave packet is often taken to have a Gaussian shape and is called a *Gaussian wave packet*. Gaussian wave packets also are used to analyze water waves. For example, a Gaussian wavefunction ψ might take the form:

$$\psi(x, t=0) = A \exp\left(-\frac{x^2}{2\sigma^2} + ik_0 x\right),$$

at some initial time $t = 0$, where the central wavelength is related to the central wave vector k_0 as $\lambda_0 = 2\pi / k_0$. It is well known from the theory of Fourier analysis, or from the Heisenberg uncertainty principle (in the case of quantum mechanics) that a narrow range of wavelengths is necessary to produce a localized wave packet, and the more localized the envelope, the larger the spread in required wavelengths. The Fourier transform of a Gaussian is itself a Gaussian. Given the Gaussian:

$$f(x) = e^{-x^2/(2\sigma^2)},$$

the Fourier transform is:

$$\tilde{f}(k) = \sigma e^{-\sigma^2 k^2/2}.$$

The Gaussian in space therefore is made up of waves:

$$f(x) = \frac{1}{\sqrt{2\pi}} \int_{-\infty}^{\infty} \tilde{f}(k) e^{ikx}\, dk \; ;$$

that is, a number of waves of wavelengths λ such that $k\lambda = 2\pi$.

The parameter σ decides the spatial spread of the Gaussian along the *x*-axis, while the Fourier transform shows a spread in wave vector *k* determined by 1/σ. That is, the smaller the extent in space, the larger the extent in *k*, and hence in $\lambda = 2\pi/k$.

Gravitational Waves

Researchers believe that gravitational waves also travel through space, although gravitational waves have never been directly detected. Not to be confused with gravity waves, gravitational waves are disturbances in the curvature of spacetime, predicted by Einstein's theory of general relativity.

WKB Method

In a nonuniform medium, in which the wavenumber *k* can depend on the location as well as the frequency, the phase term *kx* is typically replaced by the integral of *k*(*x*)*dx*, according to the WKB method. Such nonuniform travelling waves are common in many physical problems, including the mechanics of the cochlea and waves on hanging ropes.

Flood

A flood is an overflow of water that submerges land which is normally dry. The European Union (EU) Floods Directive defines a flood as a covering by water of land not normally covered by water. In the sense of "flowing water", the word may also be applied to the inflow of the tide. Flooding may occur as an overflow of water from water bodies, such as a river or lake, in which the water overtops or breaks levees, resulting in some of that water escaping its usual boundaries, or it may occur due to an accumulation of rainwater on saturated ground in an areal flood. While the size of a lake or other body of water will vary with seasonal changes in precipitation and snow melt, these changes in size are unlikely to be considered significant unless they flood property or drown domestic animals.

Floods can also occur in rivers when the flow rate exceeds the capacity of the river channel, particularly at bends or meanders in the waterway. Floods often cause damage to homes and businesses if they are in the natural flood plains of rivers. While riverine flood damage can be eliminated by moving away from rivers and other bodies of water, people have traditionally lived and worked by rivers because the land is usually flat and fertile and because rivers provide easy travel and access to commerce and industry.

Some floods develop slowly, while others such as flash floods, can develop in just a few minutes and without visible signs of rain. Additionally, floods can be local, impacting a neighbourhood or community, or very large, affecting entire river basins.

Etymology

The word "flood" comes from the Old English *flod*, a word common to Germanic languages (compare German *Flut*, Dutch *vloed* from the same root as is seen in *flow, float*; also compare with Latin *fluctus, flumen*). Deluge myths are mythical stories of a great flood sent by a deity or deities to destroy civilization as an act of divine retribution, and they are featured in the mythology of many cultures.

Principal Types and Causes

Areal (Rainfall Related)

Areal (Rainfall Related): Floods can happen on flat or low-lying areas when the ground is saturated and water either cannot run off or cannot run off quickly enough to stop accumulating. This may be followed by a river flood as water moves away from the floodplain into local rivers and streams.

Floods can also occur if water falls on an impermeable surface, such as concrete, paving or frozen ground, and cannot rapidly dissipate into the ground.

Localised heavy rain from a series of storms moving over the same area can cause areal flash flooding when the rate of rainfall exceeds the drainage capacity of the area. When this occurs on tilled fields, it can result in a muddy flood where sediments are picked up by run off and carried as suspended matter or bed load.

Riverine

River flows may rise to floods levels at different rates, from a few minutes to several weeks, depending on the type of river and the source of the increased flow.

Slow rising floods most commonly occur in large rivers with large catchment areas. The increase in flow may be the result of sustained rainfall, rapid snow melt, monsoons, or tropical cyclones. Localised flooding may be caused or exacerbated by drainage obstructions such as landslides, ice, or debris. Rapid flooding events, including flash floods, more often occur on smaller rivers, rivers with steep valleys or rivers that flow for much of their length over impermeable terrain. The cause may be localised convective precipitation (intense thunderstorms) or sudden release from an upstream impoundment created behind a dam, landslide, or glacier.

Dam-building beavers can flood low-lying urban and rural areas, occasionally causing some damage.

Estuarine and Coastal

Flooding in estuaries is commonly caused by a combination of sea tidal surges caused by winds and low barometric pressure, and they may be exacerbated by high upstream river flow. Coastal areas may be flooded by storm events at sea, resulting in waves over-topping defences or in severe cases by tsunami or tropical cyclones. A storm surge, from either a tropical cyclone or an extratropical cyclone, falls within this category.

Catastrophic

Catastrophic flooding is usually associated with major infrastructure failures such as the collapse of a dam, but they may also be caused by damage sustained in an earthquake or volcanic eruption.

Effects

Primary Effects: The primary effects of flooding include loss of life, damage to buildings and other structures, including bridges, sewerage systems, roadways, and canals.

Infrastructure damage also frequently damages power transmission and sometimes power generation, which then has knock-on effects caused by the loss of power. This includes loss of drinking water treatment and water supply, which may result in loss of drinking water or severe water contamination. It may also cause the loss of sewage disposal facilities. Lack of clean water combined with human sewage in the flood waters raises the risk of waterborne diseases, which can include typhoid, giardia, cryptosporidium, cholera and many other diseases depending upon the location of the flood.

Damage to roads and transport infrastructure may make it difficult to mobilise aid to those affected or to provide emergency health treatment. Flood waters typically inundate farm land, making the land unworkable and preventing crops from being planted or harvested, which can lead to shortages of food both for humans and farm animals. Entire harvests for a country can be lost in extreme flood circumstances. Some tree species may not survive prolonged flooding of their root systems

Secondary and Long-Term Effects

Economic hardship due to a temporary decline in tourism, rebuilding costs, or food shortages leading to price increases is a common after-effect of severe flooding. The impact on those affected may cause psychological damage to those affected, in particular where deaths, serious injuries and loss of property occur.

Flood Forecasting

Anticipating floods before they occur allows for precautions to be taken and people to be warned so that they can be prepared in advance for flooding conditions. For example, farmers can remove animals from low-lying areas and utility services can put in place emergency provisions to re-route services if needed. Emergency services can also make provisions to have enough resources available ahead of time to respond to emergencies as they occur.

In order to make the most accurate flood forecasts for waterways, it is best to have a long time-series of historical data that relates stream flows to measured past rainfall events. Coupling this historical information with real-time knowledge about volumetric capacity in catchment areas, such as spare capacity in reservoirs, ground-water levels, and the degree of saturation of area aquifers is also needed in order to make the most accurate flood forecasts.

Radar estimates of rainfall and general weather forecasting techniques are also important components of good flood forecasting. In areas where good quality data is available, the intensity and height of a flood can be predicted with fairly good accuracy and plenty of lead time. The output of a flood forecast is typically a maximum expected water level and the likely time of its arrival at key locations along a waterway, and it also may allow for the computation of the likely statistical return period of a flood. In many developed countries, urban areas at risk of flooding are protected against a 100-year flood - that is a flood that has a probability of around 63% of occurring in any 100 year period of time.

According to the U.S. National Weather Service (NWS) Northeast River Forecast Centre (RFC) in Taunton, Massachusetts, a general rule-of-thumb for flood forecasting in urban areas is that it takes at least 1 inch (25 mm) of rainfall in around an hour's time in order to start significant ponding of water on impermeable surfaces. Many NWS RFCs routinely issue Flash Flood Guidance and Headwater Guidance, which indicate the general amount of rainfall that would need to fall in a short period of time in order to cause flash flooding or flooding on larger water basins.

Control

In many countries around the world, waterways prone to floods are often carefully managed. Defences such as detention basins, levees, bunds, reservoirs, and weirs are used to prevent waterways from overflowing their banks. When these defences fail, emergency measures such as sandbags or portable inflatable tubes are often used to try and stem flooding. Coastal flooding has been addressed in portions of Europe and the Americas with coastal defences, such as sea walls, beach nourishment, and barrier islands.

In the riparian zone near rivers and streams, erosion control measures can be taken to try and slow down or reverse the natural forces that cause many waterways to meander over long periods of time. Flood controls, such as dams, can be built and maintained over time to try and reduce the occurrence and severity of floods as well. In the USA, the U.S. Army Corps of Engineers maintains a network of such flood control dams.

Benefits

Floods (in particular more frequent or smaller floods) can also bring many benefits, such as recharging ground water, making soil more fertile and increasing nutrients in some soils. Flood waters provide much needed water resources in arid and semi-arid regions where precipitation can be very unevenly distributed throughout the year. Freshwater floods particularly play an important role in maintaining ecosystems in river corridors and are a key factor in maintaining floodplain biodiversity. Flooding can spread nutrients to lakes and rivers, which can lead to increased biomass and improved fisheries for a few years. For some fish species, an inundated floodplain may form a highly suitable location for spawning with few predators and enhanced levels of nutrients or food. Fish, such as the weather fish, make use of floods in order to reach new habitats. Bird populations may also profit from the boost in food production caused by flooding.

Periodic flooding was essential to the well-being of ancient communities along the Tigris-Euphrates Rivers, the Nile River, the

Indus River, the Ganges and the Yellow River among others. The viability of hydropower, a renewable source of energy, is also higher in flood prone regions.

Computer Modelling

While flood computer modelling is a fairly recent practice, attempts to understand and manage the mechanisms at work in floodplains have been made for at least six millennia. Recent developments in computational flood modelling have enabled engineers to step away from the tried and tested "hold or break" approach and its tendency to promote overly engineered structures. Various computational flood models have been developed in recent years; either 1D models (flood levels measured in the channel) or 2D models (variable flood depths measured across the extent of a floodplain). HEC-RAS, the Hydraulic Engineering Centre model, is currently among the most popular computer models, if only because it is available free of charge. Other models such as TUFLOW combine 1D and 2D components to derive flood depths across both river channels and the entire floodplain. To date, the focus of computer modelling has primarily been on mapping tidal and fluvial flood events, but the 2007 flood events in the UK have shifted the emphasis there onto the impact of surface water flooding.

In the United States, an integrated approach to real-time hydrologic computer modelling utilises observed data from the U.S. Geological Survey (USGS), various cooperative observing networks, various automated weather sensors, the NOAA National Operational Hydrologic Remote Sensing Centre (NOHRSC), various hydroelectric companies, etc. combined with quantitative precipitation forecasts (QPF) of expected rainfall and/or snow melt to generate daily or as-needed hydrologic forecasts. The NWS also cooperates with Environment Canada on hydrologic forecasts that affect both the USA and Canada, like in the area of the Saint Lawrence Seaway.

In Myth and Religion

Flood myths (great, civilization-destroying floods as divine retribution) are widespread in many cultures and religions. As a prime example, the Genesis flood narrative plays a prominent role in Judaism and Christianity.

Chapter 6

Dimensional Analysis

In physics and all science, dimensional analysis is the practice of checking relations among physical quantities by identifying their dimensions. The dimension of any physical quantity is the combination of the basic physical dimensions that compose it. Some fundamental physical dimensions are length, mass, time, and electric charge. Other physical quantities can be expressed in terms of these fundamental physical dimensions. For example, speed has the dimension length (or distance) per unit of time, and may be measured in metres per second, miles per hour, or other units. Similarly electrical current is electrical charge per unit time (flow rate of charge) and is measured in coulombs (a unit of electrical charge) per second, or equivalently, amperes. Dimensional analysis is based on the fact that a physical law must be independent of the units used to measure the physical variables. A straightforward practical consequence is that any meaningful equation (and any inequality and inequation) must have the same dimensions on the left and right sides. Checking this is the basic way of performing dimensional analysis.

Dimensional analysis is routinely used to check the plausibility of derived equations and computations. It is also used to form reasonable hypotheses about complex physical situations that can be tested by experiment or by more developed theories of the phenomena, and to categorize types of physical quantities and units based on their relations to or dependence on other units, or their dimensions if any.

Great Principle of Similitude

The basic principle of dimensional analysis was known to Isaac Newton (1686) who referred to it as the "*Great Principle of Similitude*". James Clerk Maxwell played a major role in establishing modern use of dimensional analysis by distinguishing mass, length, and time as fundamental units, while referring to other units as derived. The 19th-century French mathematician Joseph Fourier made important contributions based on the idea that physical laws like $F = ma$ should be independent of the units employed to measure the physical variables. This led to the conclusion that meaningful laws must be homogeneous equations in their various units of measurement, a result which was eventually formalized in the Buckingham ∂ theorem. This theorem describes how every physically meaningful equation involving n variables can be equivalently rewritten as an equation of $n - m$ dimensionless parameters, where m is the rank of the dimensional matrix. Furthermore, and most importantly, it provides a method for computing these dimensionless parameters from the given variables.

A dimensional equation can have the dimensions reduced or eliminated through nondimensionalization, which begins with dimensional analysis, and involves scaling quantities by characteristic units of a system or natural units of nature. This gives insight into the fundamental properties of the system, as illustrated in the examples below.

Definition

The dimension of a physical quantity can be expressed as a product of the basic physical dimensions mass, length, time, electric charge, and absolute temperature, represented by sans-serif symbols M, L, T, Q, and Θ, respectively, each raised to a rational power.

The SI standard recommends the usage of the following dimensions and corresponding symbols: mass (M), length (L), time (T), electrical current (I), absolute temperature (Θ), amount of substance (N) and luminous intensity (J).

The term *dimension* is more abstract than *scale* unit: *mass* is a dimension, while kilograms are a scale unit (choice of standard) in the mass dimension.

As examples, the dimension of the physical quantity speed is *length/time* (L/T or LT^{-1}), and the dimension of the physical quantity force is "mass × acceleration" or "mass×(length/time)/time" (ML/T^2 or MLT^{-2}). In principle, other dimensions of physical quantity could be defined as "fundamental" (such as momentum or energy or electric current) in lieu of some of those shown above. Most physicists do not recognise temperature, Θ, as a fundamental dimension of physical quantity since it essentially expresses the energy per particle per degree of freedom, which can be expressed in terms of energy (or mass, length, and time). Still others do not recognise electric current, I, as a separate fundamental dimension of physical quantity, since it has been expressed in terms of mass, length, and time in unit systems such as the cgs system. There are also physicists that have cast doubt on the very existence of incompatible fundamental dimensions of physical quantity.

The unit of a physical quantity and its dimension are related, but not identical concepts. The units of a physical quantity are defined by convention and related to some standard; e.g., length may have units of metres, feet, inches, miles or micrometres; but any length always has a dimension of L, independent of what units are arbitrarily chosen to measure it. Two different units of the same physical quantity have conversion factors that relate them. For example: 1 in = 2.54 cm; then (2.54 cm/in) is the conversion factor, and is itself dimensionless and equal to one. Therefore multiplying by that conversion factor does not change a quantity. Dimensional symbols do not have conversion factors.

Mathematical Properties

The dimensions that can be formed from a given collection of basic physical dimensions, such as M, L, and T, form an abelian group: The identity is written as 1; $L^0 = 1$, and the inverse to L is 1/L or L^{-1}. L raised to any rational power p is a member of the group, having an inverse of L^{-p} or $1/L^p$. The operation of the group is multiplication, having the usual rules for handling exponents ($L^n \times L^m = L^{n+m}$).

This group can be described as a vector space over the rational numbers, with for example dimensional symbol $M^iL^jT^k$ corresponding to the vector (i, j, k). When physical measured quantities (be they like-dimensioned or unlike-dimensioned) are multiplied or divided by one other, their dimensional units are likewise multiplied or divided; this corresponds to addition or subtraction in the vector space. When measurable quantities are raised to a rational power, the same is done to the dimensional symbols attached to those quantities; this corresponds to scalar multiplication in the vector space.

A basis for a given vector space of dimensional symbols is called a set of fundamental units or fundamental dimensions, and all other vectors are called derived units. As in any vector space, one may choose different bases, which yields different systems of units (e.g., choosing whether the unit for charge is derived from the unit for current, or vice versa).

The group identity 1, the dimension of dimensionless quantities, corresponds to the origin in this vector space.

The set of units of the physical quantities involved in a problem correspond to a set of vectors (or a matrix). The kernel describes some number (e.g., m) of ways in which these vectors can be combined to produce a zero vector. These correspond to producing (from the measurements) a number of dimensionless quantities, $\{\pi_1,...,\pi_m\}$. (In fact these ways completely span the null subspace of another different space, of powers of the measurements.) Every possible way of multiplying (and exponentiating) together the measured quantities to produce something with the same units as some derived quantity X can be expressed in the general form

$$X = \prod_{i=1}^{m} (\pi_i)^{k_i}.$$

Consequently, every possible commensurate equation for the physics of the system can be rewritten in the form

$$f(\pi_1, \pi_2, ..., \pi_m) = 0.$$

Knowing this restriction can be a powerful tool for obtaining new insight into the system.

Mechanics

In mechanics, the dimension of any physical quantity can be expressed in terms of the fundamental dimensions (or *base dimensions*) M, L, and T – these form a 3-dimensional vector space. This is not the only possible choice, but it is the one most commonly used. For example, one might choose force, length and mass as the base dimensions (as some have done), with associated dimensions F, L, M; this corresponds to a different basis, and one may convert between these representations by a change of basis.

The choice of the base set of dimensions is, thus, partly a convention, resulting in increased utility and familiarity. It is, however, important to note that the choice of the set of dimensions cannot be chosen arbitrarily – it is not *just* a convention – because the dimensions must form a basis: they must span the space, and be linearly independent.

For example, F, L, M form a set of fundamental dimensions because they form an equivalent basis to *M, L, T:* the former can be expressed as $[F=ML/T^2]$,L,M while the latter can be expressed as M,L,$[T=(ML/F)^{1/2}]$.

On the other hand, using length, velocity and time (*L, V, T*) as base dimensions will not work well (they do not form a set of fundamental dimensions), for two reasons:

- There is no way to obtain mass – or anything derived from it, such as force – without introducing another base dimension (thus these do not *span the space*).
- Velocity, being derived from length and time (V=L/T), is redundant (the set is not *linearly independent*).

Other Fields of Physics and Chemistry

Depending on the field of physics, it may be advantageous to choose one or another extended set of dimensional symbols. In electromagnetism, for example, it may be useful to use dimensions of M, L, T, and Q, where Q represents quantity of electric charge. In thermodynamics, the base set of dimensions is often extended to include a dimension for temperature, Θ. In chemistry the number of moles of substance (loosely, but not precisely, related to the number of molecules or atoms) is often involved and a dimension for this is used as well. In the interaction of relativistic plasma with strong laser pulses a dimensionless relativistic similarity parameter connected with the symmetry properties of the collisionless Vlasov equation is constructed from the plasma electron and critical densities in addition to the electromagnetic vector potential. The choice of the dimensions or even the number of dimensions to be used in different fields of physics is to some extent arbitrary, but consistency in use and ease of communications are very important.

Commensurability

The most basic rule of dimensional analysis is that of dimensional homogeneity, first formulated by Newton, who called it the "great principle of similitude":

Only *commensurable* quantities (quantities with the same dimensions) may be *compared, equated, added,* or *subtracted.*

However, as stated above, the dimensions form a multiplicative group; consequently:

One may take *ratios* of *incommensurable* quantities (quantities with different dimensions), and *multiply* or *divide* them.

For example, it makes no sense to ask if 1 hour is more, the same, or less than 1 kilometre, as these have different dimensions, nor to add 1 hour to 1 kilometre. On the other hand, if an object travels

100 km in 2 hours, one may divide these and conclude that the object's average speed was 50 km/h.

The rule implies that in a physically meaningful *expression* only quantities of the same dimension can be added, subtracted, or compared. For example, if m_{man}, m_{rat} and L_{man} denote, respectively, the mass of some man, the mass of a rat and the length of that man, the dimensionally homogeneous expression $m_{man} + m_{rat}$ is meaningful, but the heterogeneous expression $m_{man} + L_{man}$ is meaningless. However, m_{man}/L^2_{man} is fine. Thus, dimensional analysis may be used as a sanity check of physical equations: the two sides of any equation must be commensurable or have the same dimensions.

Even when two physical quantities have identical dimensions, it may nevertheless be meaningless to compare or add them. For example, although torque and energy share the dimension ML^2/T^2, they are fundamentally different physical quantities. To compare, add, or subtract quantities with the same dimensions but expressed in different units, the standard procedure is to first convert them all to the same units. For example, to compare 32 metres with 35 yards, use 1 yard = 0.9144 m to convert 35 yards to 32.004 m.

Polynomials and Transcendental Functions

Scalar arguments to transcendental functions such as exponential, trigonometric and logarithmic functions, or to inhomogeneous polynomials, must be dimensionless quantities. (Note: this requirement is somewhat relaxed in Siano's orientational analysis described below, in which the square of certain dimensioned quantities are dimensionless.) While most mathematical identities about dimensionless numbers translate in a straightforward manner to dimensional quantities, care must be taken with logarithms of ratios: the identity log(a/b) = log a " log b, where the logarithm is taken in any base, holds for dimensionless numbers a and b, but it does *not* hold if a and b are dimensional, because in this case the left-hand side is well-defined but the right-hand side is not.

Similarly, while one can evaluate monomials (x^n) of dimensional quantities, one cannot evaluate polynomials of mixed degree with dimensionless coefficients on dimensional quantities: for x^2, the expression $(3\ m)^2 = 9\ m^2$ makes sense (as an area), while for $x^2 + x$, the expression $(3\ m)^2 + 3\ m = 9\ m^2 + 3\ m$ does not make sense.

However, polynomials of mixed degree can make sense if the coefficients are suitably chosen physical quantities that are not dimensionless. For example,

$$\frac{1}{2}\cdot\left(-32\frac{\text{foot}}{\text{second}^2}\right)\cdot t^2+\left(500\frac{\text{foot}}{\text{second}}\right)\cdot t.$$

This is the height to which an object rises in time t if the acceleration of gravity is 32 feet per second per second and the initial upward speed is 500 feet per second. It is not even necessary for t to be in *seconds*. For example, suppose $t = 0.01$ minutes. Then the first term would be

$$\frac{1}{2}\cdot\left(-32\frac{\text{foot}}{\text{second}^2}\right)\cdot(0.01\ \text{minute})^2$$

$$=\frac{1}{2}\cdot -32\cdot(0.01^2)\left(\frac{\text{minute}}{\text{second}}\right)^2\cdot\text{foot}$$

$$=\frac{1}{2}\cdot -32\cdot(0.01^2)\cdot 60^2\cdot\text{foot}.$$

Incorporating Units

The value of a dimensional physical quantity Z is written as the product of a unit $[Z]$ within the dimension and a dimensionless numerical factor, n.

$$Z = n\times[Z] = n[Z]$$

In a strict sense, when like-dimensioned quantities are added or subtracted or compared, these dimensioned quantities must be expressed in consistent units so that the numerical values of these quantities may be directly added or subtracted. But, in concept, there is no problem adding quantities of the same dimension expressed in different units. For example, 1 metre added to 1 foot *is* a length, but it would not be correct to add 1 to 1 to get the result. A conversion factor, which is a ratio of like-dimensioned quantities and is equal to the dimensionless unity, is needed:

$$1\ \text{ft} = 0.3048\ \text{m} \text{ is identical to } 1=\frac{0.3048\ \text{m}}{1\ \text{ft}}.$$

The factor $0.3048\frac{\text{m}}{\text{ft}}$ is identical to the dimensionless 1, so multiplying by this conversion factor changes nothing. Then when adding two quantities of like dimension, but expressed in different units, the appropriate conversion factor, which is essentially the dimensionless 1, is used to convert the quantities to identical units so that their numerical values can be added or subtracted. Only in this manner is it meaningful to speak of adding like-dimensioned quantities of differing units.

Position vs Displacement

Some discussions of dimensional analysis implicitly describe all quantities as mathematical vectors. (In mathematics scalars are considered a special case of vectors; the emphasis here is that vectors are closed under addition, subtraction, and scalar multiplication, and permit scalar division.). This assumes an implicit point of reference – an origin. While this is useful and often perfectly adequate, allowing many important errors to be caught, it can fail to model certain aspects of physics. A more rigorous approach requires distinguishing between position and displacement (or moment in time versus duration, or absolute temperature versus temperature change).

Consider points on a line, each with a position with respect to a given origin, and distances among them. Positions and displacements all have units of length, but their meaning is not interchangeable:

- adding two displacements should yield a new displacement (walking ten paces then twenty paces gets you thirty paces forward),
- adding a displacement to a position should yield a new position (walking one block down the street from an intersection gets you to the next intersection),
- subtracting two positions should yield a displacement,
- but one may *not* add two positions.

This illustrates the subtle distinction between *affine* quantities (ones modelled by an affine space, such as position) and *vector* quantities (ones modelled by a vector space, such as displacement).

- Vector quantities may be added to each other, yielding a new vector quantity, and a vector quantity may be added to a suitable affine quantity (a vector space *acts on* an affine space), yielding a new affine quantity.
- Affine quantities cannot be added, but may be subtracted, yielding *relative* quantities which are vectors, and these *relative differences* may then be added to each other or to an affine quantity.

Properly then, positions have dimension of *affine* length, while displacements have dimension of *vector* length. To assign a number to an *affine* unit, one must not only choose a unit of measurement, but also a point of reference, while to assign a number to a *vector* unit only requires a unit of measurement.

Thus some physical quantities are better modelled by vectorial quantities while others tend to require affine representation, and the distinction is reflected in their dimensional analysis.

This distinction is particularly important in the case of temperature, for which the numeric value of absolute zero is not the origin 0 in some scales. For absolute zero,

0 K = –273.15 °C = –459.67 °F = 0 °R,

but for temperature differences,

1 K = 1 °C ≠ 1 °F = 1 °R.

(Here °R refers to the Rankine scale, not the Réaumur scale). Unit conversion for temperature differences is simply a matter of multiplying by, e.g., 1 °F / 1 K. But because some of these scales have origins that do not correspond to absolute zero, conversion from one temperature scale to another requires accounting for that. As a result, simple dimensional analysis can lead to errors if it is ambiguous whether 1 K means the absolute temperature equal to –272.15 °C, or the temperature difference equal to 1 °C.

Orientation and Frame of Reference

Similar to the issue of a point of reference is the issue of orientation: a displacement in 2 or 3 dimensions is not just a length, but is a length together with a *direction*. (This issue does not arise in 1 dimension, or rather is equivalent to the distinction between positive and negative.) Thus, to compare or combine two dimensional quantities in a multi-dimensional space, one also needs an orientation: they need to be compared to a frame of reference.

This leads to the extensions discussed below, namely Huntley's directed dimensions and Siano's orientational analysis.

Other Uses

Dimensional analysis is also used to derive relationships between the physical quantities that are involved in a particular phenomenon that one wishes to understand and characterize. It was used for the first time (Pesic 2005) in this way in 1872 by Lord Rayleigh, who was trying to understand why the sky is blue. Rayleigh first published the technique is his book "theory of sound" from 1877.

Examples

A simple example: period of a harmonic oscillator:

What is the period of oscillation T of a mass m attached to an ideal linear spring with spring constant k suspended in gravity of strength g? That period is the solution for T of some dimensionless equation in the variables T, m, k, and g. The four quantities have the following dimensions: T [T]; m [M]; k [M/T^2]; and g [L/T^2]. From these we can form only one dimensionless product of powers of our chosen variables, $G_1 = T^2k/m$ [$T^2 \cdot M/T^2 / M = 1$], and putting $G_1 = C$ for some dimensionless constant C gives the dimensionless equation sought. The dimensionless product of powers of variables is sometimes referred to as a dimensionless group of variables; here the term "group" means "collection" rather than mathematical group. They are often called dimensionless numbers as well.

Note that the variable g does not occur in the group. It is easy to see that it is impossible to form a dimensionless product of powers that combines g with k, m, and T, because g is the only quantity that involves the dimension L. This implies that in this problem the g is irrelevant. Dimensional analysis can sometimes yield strong statements about the *irrelevance* of some quantities in a problem, or the need for additional parameters. If we have chosen enough variables to properly describe the problem, then from this argument we can conclude that the period of the mass on the spring is independent of g: it is the same on the earth or the moon. The equation demonstrating the existence of a product of powers for our problem can be written in an entirely equivalent way: $T = \kappa\sqrt{m/k}$, for some dimensionless constant κ (equal to $\sqrt{C}$ from the original dimensionless equation).

When faced with a case where dimensional analysis rejects a variable (g, here) that one intuitively expects to belong in a physical description of the situation, another possibility is that the rejected variable is in fact relevant, but that some other relevant variable has been omitted, which might combine with the rejected variable to form a dimensionless quantity. That is, however, not the case here.

When dimensional analysis yields only one dimensionless group, as here, there are no unknown functions, and the solution is said to be "complete" – although it still may involve unknown dimensionless constants, such as κ.

A More Complex Example: Energy of a Vibrating Wire

Consider the case of a vibrating wire of length ℓ (L) vibrating with an amplitude A (L). The wire has a linear density ρ (M/L) and is under tension s (ML/T^2), and we want to know the energy E (ML^2/T^2) in the wire. Let π_1 and π_2 be two dimensionless products of powers of the variables chosen, given by

$$\pi_1 = E / As$$
$$\pi_2 = \ell / A.$$

The linear density of the wire is not involved. The two groups found can be combined into an equivalent form as an equation

$$F(E / As, \ell / A) = 0,$$

where F is some unknown function, or, equivalently as

$$E = Asf(\ell / A),$$

where f is some other unknown function. Here the unknown function implies that our solution is now incomplete, but dimensional analysis has given us something that may not have been obvious: the energy is proportional to the first power of the tension. Barring further analytical analysis, we might proceed to experiments to discover the form for the unknown function f. But our experiments are simpler than in the absence of dimensional analysis. We'd perform none to verify that the energy is proportional to the tension. Or perhaps we might guess that the energy is proportional to ℓ, and so infer that $E = \ell s$. The power of dimensional analysis as an aid to experiment and forming hypotheses becomes evident.

The power of dimensional analysis really becomes apparent when it is applied to situations, unlike those given above, that are more complicated, the set of variables involved are not apparent, and the underlying equations hopelessly complex. Consider, for example, a small pebble sitting on the bed of a river. If the river flows fast enough, it will actually raise the pebble and cause it to flow along with the water. At what critical velocity will this occur? Sorting out the guessed variables is not so easy as before. But dimensional analysis can be a powerful aid in understanding problems like this, and is usually the very first tool to be applied to complex problems where the underlying equations and constraints are poorly understood. In such cases, the answer may depend on a dimensionless number such as the Reynolds number, which may be interpreted by dimensional analysis.

Extensions

Huntley's Extension: Directed Dimensions: Huntley (Huntley 1967) has pointed out that it is sometimes productive to refine our concept of dimension. Two possible refinements are:

- The magnitude of the components of a vector are to be considered dimensionally distinct. For example, rather than an undifferentiated length unit L, we may have L_x represent length in the x direction, and so forth. This requirement stems ultimately from the requirement that each component of a physically meaningful equation (scalar, vector, or tensor) must be dimensionally consistent.

- Mass as a measure of quantity is to be considered dimensionally distinct from mass as a measure of inertia.

As an example of the usefulness of the first refinement, suppose we wish to calculate the distance a cannonball travels when fired with a vertical velocity component V_y and a horizontal velocity component V_x, assuming it is fired on a flat surface. Assuming no use of directed lengths, the quantities of interest are then V_x, V_y, both dimensioned as L/T, R, the distance travelled, having dimension L, and g the downward acceleration of gravity, with dimension L/T^2

With these four quantities, we may conclude that the equation for the range R may be written:

$$R \propto V_x^a V_y^b g^c.$$

Or dimensionally

$$L = (L/T)^{a+b}(L/T^2)^c$$

from which we may deduce that $a+b+c=1$ and $a+b+2c=0$, which leaves one exponent undetermined. This is to be expected since we have two fundamental quantities L and T and four parameters, with one equation.

If, however, we use directed length dimensions, then V_x will be dimensioned as L_x/T, V_y as L_y/T, R as L_x and g as L_y/T^2. The dimensional equation becomes:

$$L_x = (L_x/T)^a (L_y/T)^b (L_y/T^2)^c$$

and we may solve completely as $a=1$, $b=1$ and $c=-1$. The increase in deductive power gained by the use of directed length dimensions is apparent.

In a similar manner, it is sometimes found useful (e.g., in fluid mechanics and thermodynamics) to distinguish between mass as a measure of inertia (inertial mass), and mass as a measure of quantity (substantial mass). For example, consider the derivation of Poiseuille's Law. We wish to find the rate of mass flow of a viscous fluid through a circular pipe. Without drawing distinctions between inertial and substantial mass we may choose as the relevant variables

- $\dot{m}$ the mass flow rate with dimensions M / T
- p_x the pressure gradient along the pipe with dimensions M / L^2T^2
- ρ the density with dimensions M / L^3
- η the dynamic fluid viscosity with dimensions M / LT
- r the radius of the pipe with dimensions L

There are three fundamental variables so the above five equations will yield two dimensionless variables which we may take to be $\pi_1 = \dot{m} / \eta r$ and $\pi_2 = p_x \rho r^5 / \dot{m}^2$ and we may express the dimensional equation as

$$C = \pi_1 \pi_2^a = \left(\frac{\dot{m}}{\eta r}\right)\left(\frac{p_x \rho r^5}{\dot{m}^2}\right)^a$$

where C and a are undetermined constants. If we draw a distinction between inertial mass with dimensions M_i and substantial mass with dimensions M_s, then mass flow rate and density will use substantial mass as the mass parameter, while the pressure gradient and coefficient of viscosity will use inertial mass. We now have four fundamental parameters, and one dimensionless constant, so that the dimensional equation may be written:

$$C = \frac{p_x \rho r^4}{\eta \dot{m}}$$

where now only C is an undetermined constant (found to be equal to $\pi / 8$ by methods outside of dimensional analysis). This equation may be solved for the mass flow rate to yield Poiseuille's law.

Siano's Extension: Orientational Analysis

Huntley's extension has some serious drawbacks:

- It does not deal well with vector equations involving the *cross product,*
- nor does it handle well the use of *angles* as physical variables.

It also is often quite difficult to assign the L, L_x, L_y, L_z, symbols to the physical variables involved in the problem of interest. He invokes a procedure that involves the "symmetry" of the physical problem. This is often very difficult to apply reliably: It is unclear as to what parts of the problem that the notion of "symmetry" is being invoked. Is it the symmetry of the physical body that forces are acting upon, or to the points, lines or areas at which forces are being applied?

What if more than one body is involved with different symmetries? Consider the spherical bubble attached to a cylindrical tube, where one wants the flow rate of air as a function of the pressure difference in the two parts. What are the Huntley extended dimensions of the viscosity of the air contained in the connected parts? What are the extended dimensions of the pressure of the two parts? Are they the same or different? These difficulties are responsible for the limited application of Huntley's addition to real problems.

Angles are, by convention, considered to be dimensionless variables, and so the use of angles as physical variables in dimensional analysis can give less meaningful results. As an example, consider the projectile problem mentioned above. Suppose that, instead of the x- and y-components of the initial velocity, we had chosen the magnitude of the velocity v and the angle θ at which the projectile was fired. The angle is, by convention, considered to be dimensionless, and the magnitude of a vector has no directional quality, so that no dimensionless variable can be composed of the four variables g, v, R, and θ. Conventional analysis will correctly give the powers of g and v, but will give no information concerning the dimensionless angle θ.

Siano (1985-I, 1985-II) has suggested that the directed dimensions of Huntley be replaced by using *orientational symbols* 1_x 1_y 1_z to denote vector directions, and an orientationless symbol 1_0. Thus, Huntley's 1_x becomes $L\,1_x$ with L specifying the dimension of length, and 1_x specifying the orientation. Siano further shows that the orientational symbols have an algebra of their own. Along with the requirement that $1_i^{-1} = 1_i$, the following multiplication table for the orientation symbols results:

	1_0	1_x	1_y	1_z
1_0	1_0	1_x	1_y	1_z
1_x	1_x	1_0	1_z	1_y
1_y	1_y	1_z	1_0	1_x
1_z	1_z	1_y	1_x	1_0

Note that the orientational symbols form a group (the Klein four-group or "Viergruppe"). In this system, scalars always have the same orientation as the identity element, independent of the "symmetry of the problem." Physical quantities that are vectors have the orientation expected: a force or a velocity in the z-direction has the orientation of 1_z. For angles, consider an angle θ that lies in the z-plane. Form a right triangle in the z plane with θ being one of the acute angles.

The side of the right triangle adjacent to the angle then has an orientation 1_x and the side opposite has an orientation 1_y. Then, since $\tan(\theta) = 1_y/1_x = \theta + \ldots$ we conclude that an angle in the xy plane must have an orientation $1_y/1_x = 1_z$, which is not unreasonable. Analogous reasoning forces the conclusion that $\sin(\theta)$ has orientation 1_z while $\cos(\theta)$ has orientation 1_0. These are different, so one concludes (correctly), for example, that there are no solutions of physical equations that are of the form $a\cos(\theta)+b\sin(\theta)$, where a and b are real scalars. Note that an expression such as $\sin(\theta+\pi/2)=\cos(\theta)$ is not dimensionally inconsistent since it is a special case of the sum of angles formula and should properly be written:

$$\sin(a1_z + b1_z) = \sin(a1_z)\cos(b1_z) + \sin(b1_z)\cos(a1_z)$$

which for $a=\theta$ and $b=\pi/2$ yields $\sin(\theta 1_z + (\pi/2)1_z) = 1_z\cos(\theta 1_z)$. Physical quantities may be expressed as complex numbers (e.g. $e^{i\theta}$) which imply that the complex quantity i has an orientation equal to that of the angle it is associated with (1_z in the above example).

The assignment of orientational symbols to physical quantities and the requirement that physical equations be orientationally homogeneous can actually be used in a way that is similar to dimensional analysis to derive a little more information about acceptable solutions of physical problems. In this approach one sets up the dimensional equation and solves it as far as one can. If the lowest power of a physical variable is fractional, both sides of the solution is raised to a power such that all powers are integral. This puts it into "normal form". The orientational equation is then solved to give a more restrictive condition on the unknown powers of the orientational symbols, arriving at a solution that is more complete than the one that dimensional analysis alone gives. Often the added information is that one of the powers of a certain variable is even or odd.

As an example, for the projectile problem, using orientational symbols, θ, being in the xy-plane will thus have dimension 1_z and the range of the projectile R will be of the form:

$$R = g^a v^b \theta^c \text{ which means } L1_x \sim \left(\frac{L1_y}{T^2}\right)^a \left(\frac{L}{T}\right)^b 1_z^c.$$

Dimensional homogeneity will now correctly yield $a = -1$ and $b = 2$, and orientational homogeneity requires that c be an odd integer. In fact the required function of theta will be $\sin(\theta)\cos(\theta)$ which is a series of odd powers of θ.

It is seen that the Taylor series of sin(θ) and cos(θ) are orientationally homogeneous using the above multiplication table, while expressions like cos(θ) + sin(θ) and exp(θ) are not, and are (correctly) deemed unphysical.

It should be clear that the multiplication rule used for the orientational symbols is not the same as that for the cross product of two vectors. The cross product of two identical vectors is zero, while the product of two identical orientational symbols is the identity element.

Percentages and Derivatives

Percentages are dimensionless quantities, since they are ratios of two quantities with the same dimensions.

Derivatives with respect to a quantity add the dimensions of the variable one is differentiating with respect to on the denominator. Thus:

- position (x) has units of L (Length);
- derivative of position with respect to time (dx/dt, velocity) has units of L/T – Length from position, Time from the derivative;
- the second derivative (d^2x/dt^2, acceleration) has units of L/T^2.

In economics, one distinguishes between stocks and flows: a stock has units of "units" (say, widgets or dollars), while a flow is a derivative of a stock, and has units of "units/time" (say, dollars/year).

In some contexts, dimensional quantities are expressed as dimensionless quantities or percentages by omitting some dimensions. For example, Debt to GDP ratios are generally expressed as percentages: total debt outstanding (dimension of Currency) divided by annual GDP (dimension of Currency) – but one may argue that in comparing a stock to a flow, annual GDP should have dimensions of Currency/Time (Dollars/Year, for instance), and thus Debt to GDP should have units of years.

Dimensionless Concepts

Constants: The dimensionless constants that arise in the results obtained, such as the C in the Poiseuille's Law problem and the k in the spring problems discussed above come from a more detailed analysis of the underlying physics, and often arises from integrating some differential equation. Dimensional analysis itself has little to say about these constants, but it is useful to know that they very often have a magnitude of order unity. This observation can allow one to sometimes make "back of the envelope" calculations about the phenomenon of interest, and therefore be able to more efficiently design experiments to measure it, or to judge whether it is important, etc.

Formalisms

Paradoxically, dimensional analysis can be a useful tool even if all the parameters in the underlying theory are dimensionless, e.g., lattice models such as the Ising model can be used to study phase transitions and critical phenomena. Such models can be formulated in a purely dimensionless way. As we approach the critical point closer and closer, the distance over which the variables in the lattice model are correlated (the so-called correlation length, ξ) becomes larger and larger. Now, the correlation length is the relevant length scale related to critical phenomena, so one can, e.g., surmize on "dimensional grounds" that the non-analytical part of the free energy per lattice site should be $\sim 1/\xi^d$ where d is the dimension of the lattice.

It has been argued by some physicists, e.g., Michael Duff, that the laws of physics are inherently dimensionless. The fact that we have assigned incompatible dimensions to Length, Time and Mass is, according to this point of view, just a matter of convention, borne out of the fact that before the advent of modern physics, there was no way to relate mass, length, and time to each other. The three independent dimensionful constants: c, ‘, and G, in the fundamental equations of physics must then be seen as mere conversion factors to convert Mass, Time and Length into each other. Just as in the case of critical properties of lattice models, one can recover the results of dimensional analysis in the appropriate scaling limit; e.g., dimensional analysis in mechanics can be derived by reinserting the constants ‘, c, and G (but we can now consider them to be dimensionless) and demanding that a nonsingular relation between quantities exists in the limit $c \to \infty$, $\hbar \to 0$ and $G \to 0$. In problems involving a gravitational field the latter limit should be taken such that the field stays finite.

Finance, Economics, and Accounting

In finance, economics, and accounting, dimensional analysis is most commonly referred to in terms of the distinction between stocks and flows. More generally, dimensional analysis is used in interpreting various financial ratios, economics ratios, and accounting ratios.

- For example, the P/E ratio has dimensions of time (units of years), and can be interpreted as "years of earnings to earn the price paid."

- In economics, debt-to-GDP ratio also has units of years (debt has units of currency, GDP has units of currency/year).
- More surprisingly, bond duration also has units of years, which can be shown by dimensional analysis, but takes some financial intuition to understand.
- Velocity of money has units of 1/Years (GDP/Money supply has units of Currency/Year over Currency): how often a unit of currency circulates per year.
- Interest rates are often expressed as a percentage, but more properly percent per annum, which has dimensions of 1/Years.

Critics of mainstream economics, notably including adherents of Austrian economics, have claimed that it lacks dimensional consistency.

Dimensional Equivalences

Following are tables of commonly occurring expressions in physics, related to the dimensions of energy, momentum, and force.

SI Units

Energy E *[M] [L]^2[T]$^{-2}$*	*Expression*	*Nomenclature*
Mechanical	Fd	F = force, d = distance
	$S / t \equiv Pt$	S = action, t = time, P = power
	$mv^2 \equiv pv \equiv p^2 / m$	m = mass, v = velocity, p = momentum
	$I\omega^2 \equiv L\omega \equiv L^2 / I$	L = angular momentum, I = moment of inertia, ω = angular velocity
Thermal	$pV \equiv nRT \equiv k_BT \equiv TS$	p = pressure, T = temperature, S = entropy, k_B = boltzmann constant, R = gas constant
Waves	$IAt \equiv SAt$	I = wave intensity, S = Poynting vector
Electromagnetic	$q\phi$	q = electric charge, ϕ = electric potential (for changes this is voltage)
	$\epsilon E^2V \equiv B^2V / \mu$	E = electric field, B = magnetic field, ε= permittivity, μ = permeability, V = 3d volume
	$pE \equiv mB \equiv IA$	p = electric dipole moment, m = magnetic moment, A = area (bounded by a current loop), I = electric current in loop

Momentum p *[M] [L] [T]$^{-1}$*	*Expression*	*Nomenclature*
Mechanical	$mv \equiv Ft$	m = mass, v = velocity, F = force, t = time
	$S / r \equiv L / r$	S = action, L = angular momentum, r = displacement
Thermal	$m\sqrt{\langle v^2 \rangle}$	$\sqrt{\langle v^2 \rangle}$ = root mean square velocity, m = mass (of a molecule)
Waves	ρVv	ρ = mass density, V = 3d volume, v phase velocity,
Electromagnetic	qA	A = magnetic vector potential

Force F [M] [L] [T]⁻²	**Expression**	**Nomenclature**
Mechanical	$ma \equiv p/t$	m = mass, a = acceleration
Thermal	$T\delta S/\delta r$	S entropy, T = temperature, r = displacement (see entropic force)
Waves	$\rho V v$	ρ = mass density, V = 3d volume, v phase velocity,
Electromagnetic	$Eq \equiv Bqv$	E = electric field, B = magnetic field, v = velocity, q = charge

Natural Units

If $c = \hbar = 1$, where c = luminal speed and $\hbar$ = Planck's reduced constant, and a suitable fixed unit of energy is chosen, then all quantities of length L, mass M and time T can be expressed (dimensionally) as a power of energy E, because length, mass and time can be expressed using speed v, action S, and energy E:[12]

$$M = E/v^2, \quad L = Sv/E, \quad t = S/E$$

though speed and action are dimensionless ($v = c = 1$ and $S = \hbar = 1$) – so the only remaining quantity with dimension is energy. In terms of powers of dimensions:

$$E^n = M^p L^q t^r = E^{p-q-r}$$

This particularly useful in particle physics and high energy physics, in which case the energy unit is the electron volt (eV). Dimensional checks and estimates become very simple in this system.

However, if electric charges and currents are involved, another unit to be fixed is for electric charge, normally the electron charge e though other choices are possible.

Quantity	**p,q,r powers of energy**			**n power of energy**
	p	q	r	n
Action S	1	2	–1	0
Speed v	0	1	–1	0
Mass M	1	0	0	1
Length L	0	1	0	–1
Time t	0	0	1	–1
Momentum p	1	1	–1	1
Energy E	1	2	–2	1

Theoretical Physics

Theoretical physics is a branch of physics which employs mathematical models and abstractions of physical objects and systems to rationalize, explain and predict natural phenomena. This is in contrast to experimental physics, which uses experimental tools to probe these phenomena.

The advancement of science depends in general on the interplay between experimental studies and theory. In some cases, theoretical physics adheres to standards of mathematical rigor while giving little weight to experiments and observations. For example, while developing special relativity, Albert Einstein was concerned with the Lorentz transformation which left Maxwell's equations invariant, but was apparently uninterested in the Michelson–Morley experiment on Earth's drift through a luminiferous ether. On the other hand, Einstein was awarded the Nobel Prize for explaining the photoelectric effect, previously an experimental result lacking a theoretical formulation.

Overview

A physical theory is a model of physical events. It is judged by the extent to which its predictions agree with empirical observations. The quality of a physical theory is also judged on its ability to make new predictions which can be verified by new observations. A physical theory differs from a mathematical theorem in that while both are based on some form of axioms, judgment of mathematical applicability is not based on agreement with any experimental results.

$$\mathrm{Ric} = kg$$

The equations for an Einstein manifold, used in general relativity to describe the curvature of spacetime

A physical theory involves one or more relationships between various measurable quantities. Archimedes realised that a ship floats by displacing its mass of water, Pythagoras understood the relation between the length of a vibrating string and the musical tone it produces. Other examples include entropy as a measure of the uncertainty regarding the positions and motions of unseen particles and the quantum mechanical idea that (action and) energy are not continuously variable.

Theoreticial physics consists of several different approaches. In this regard, theoretical particle physics forms a good example. For instance: "phenomenologists" might employ (semi-) empirical formulas to agree with experimental results, often without deep physical understanding. "Modellers" (also called "model-builders") often appear much like phenomenologists, but try to model speculative theories that have certain desirable features (rather than on experimental data), or apply the techniques of mathematical modelling to physics problems. Some attempt to create approximate theories, called *effective theories*, because fully developed theories may be regarded as unsolvable or too complicated. Other theorists may try to unify, formalise, reinterpret or generalise extant theories, or create completely new ones altogether. Sometimes the vision provided by pure mathematical systems can provide clues to how a physical system might be modelled; e.g., the notion, due to Riemann and others, that space itself might be curved. Theoretical problems that need computational investigation are often the concern of computational physics.

Theoretical advances may consist in setting aside old, incorrect paradigms (e.g., burning consists of evolving phlogiston, or astronomical bodies revolve around the Earth) or may be an alternative model that provides answers that are more accurate or that can be more widely applied. In the latter case, a correspondence principle will be required to recover the previously known result.

Physical theories become accepted if they are able to make correct predictions and no (or few) incorrect ones. The theory should have, at least as a secondary objective, a certain economy and elegance (compare to mathematical beauty), a notion sometimes called "Occam's razor" after the 13th-century English philosopher William of Occam (or Ockham), in which the simpler of two theories that describe the same matter just as adequately is preferred (but conceptual simplicity may mean mathematical complexity). They are also more likely to be accepted if they connect a wide range of phenomena. Testing the consequences of a theory is part of the scientific method.

Physical theories can be grouped into three categories: *mainstream theories*, *proposed theories* and *fringe theories*.

History

Theoretical physics began at least 2,300 years ago, under the Pre-socratic philosophy, and continued by Plato and Aristotle, whose views held sway for a millennium. During the rise of medieval universities, the only acknowledged intellectual disciplines were the seven liberal arts of the *Trivium* like grammar, logic, and rhetoric and of the *Quadrivium* like arithmetic, geometry, music and astronomy. As the concepts of matter, energy, space, time and causality slowly began to acquire the form we know today, other sciences spun off from the rubric of natural philosophy. During the Middle Ages and Renaissance, the concept of experimental science, the counterpoint to theory, began with scholars such as Ibn al-Haytham and Francis Bacon. The modern era of theory began perhaps with the Copernican paradigm shift in astronomy, soon followed by Johannes Kepler's expressions for planetary orbits, which summarized the meticulous observations of Tycho Brahe.

The great push toward the modern concept of explanation started with Galileo, one of the few physicists who was both a consummate theoretician and a great experimentalist. The analytic geometry and mechanics of Descartes were incorporated into the calculus and mechanics of Isaac Newton, another theoretician/experimentalist of the highest order, writing Principia Mathematica. Joseph-Louis Lagrange, Leonhard Euler and William Rowan Hamilton would extend the theory of classical mechanics considerably. Each of these individuals picked up the interactive intertwining of mathematics and physics begun two millennia earlier by Pythagoras. Among the great conceptual achievements of the 19th and 20th centuries were the consolidation of the idea of energy by the inclusion of heat, then electricity and magnetism and light, and finally mass. The laws of thermodynamics, and most importantly the introduction of the singular concept of entropy began to provide a macroscopic explanation for the properties of matter.

The pillars of modern physics, and perhaps the most revolutionary theories in the history of physics, have been relativity theory and quantum mechanics. Newtonian mechanics was subsumed under special relativity and Newton's gravity was given a kinematic explanation by general relativity. Quantum mechanics led to an understanding of blackbody radiation and of anomalies in the specific heats of solids — and finally to an understanding of the internal structures of atoms and molecules. All of these achievements depended on the theoretical physics as a moving force both to suggest experiments and to consolidate results — often by ingenious application of existing mathematics, or, as in the case of Descartes and Newton (with Leibniz), by inventing new mathematics. Fourier's studies of heat conduction led to a new branch of mathematics: infinite, orthogonal series.

Modern theoretical physics attempts to unify theories and explain phenomena in further attempts to understand the Universe, from the cosmological to the elementary particle scale. Where experimentation cannot be done, theoretical physics still tries to advance through the use of mathematical models.

Chapter 7

Mainstream Theories

Mainstream theories (sometimes referred to as *central theories*) are the body of knowledge of both factual and scientific views and possess a usual scientific quality of the tests of repeatability, consistency with existing well-established science and experimentation. There do exist mainstream theories that are generally accepted theories based solely upon their effects explaining a wide variety of data, although the detection, explanation, and possible composition are subjects of debate.

Examples

- Black hole thermodynamics
- Classical mechanics
- Condensed matter physics
- Conservation of energy
- Dark Energy
- Dark matter
- Dynamics
- Electromagnetism
- Field theory
- Fluid dynamics
- General relativity
- Molecular modelling
- Particle physics
- Physical cosmology
- Quantum chromodynamics

- Quantum computers
- Quantum electrochemistry
- Quantum electrodynamics
- Quantum field theory
- Quantum information theory
- Quantum mechanics
- Quantum Gravity
- Solid mechanics
- Solid state physics or Condensed Matter Physics and the electronic structure of materials
- Special relativity
- Standard Model
- Statistical mechanics
- Thermodynamics

Proposed Theories

The proposed theories of physics are usually relatively new theories which deal with the study of physics which include scientific approaches, means for determining the validity of models and new types of reasoning used to arrive at the theory. However, some proposed theories include theories that have been around for decades and have eluded methods of discovery and testing. Proposed theories can include fringe theories in the process of becoming established (and, sometimes, gaining wider acceptance). Proposed theories usually have not been tested.

Examples

- Causal Sets
- Dark energy or Einstein's Cosmological Constant
- Einstein–Rosen Bridge
- Emergence
- Grand unification theory
- Loop quantum gravity
- M-theory
- String theory
- Supersymmetry
- Theory of everything
- Unparticle physics

Fringe Theories

Fringe theories include any new area of scientific endeavour in the process of becoming established and some proposed theories. It can include speculative sciences. This includes physics fields and physical theories presented in accordance with known evidence, and a body of associated predictions have been made according to that theory.

Some fringe theories go on to become a widely accepted part of physics. Other fringe theories end up being disproven. Some fringe theories are a form of protoscience and others are a form of pseudoscience. The falsification of the original theory sometimes leads to reformulation of the theory.

Examples

- Biefeld Brown electrogravity
- Dynamic theory of gravity
- Luminiferous aether
- Orgone

Thought Experiments vs Real Experiments

"Thought" experiments are situations created in one's mind, asking a question akin to "suppose you are in this situation, assuming such is true, what would follow?". They are usually created to investigate phenomena that are not readily experienced in everyday situations. Famous examples of such thought experiments are Schrödinger's cat, the EPR thought experiment, simple illustrations of time dilation, and so on. These usually lead to real experiments designed to verify that the conclusion (and therefore the assumptions) of the thought experiments are correct. The EPR thought experiment led to the Bell inequalities, which were then tested to various degrees of rigor, leading to the acceptance of the current formulation of quantum mechanics and probabilism as a working hypothesis.

Pipeline Transport

Pipeline transport is the transportation of goods through a pipe. Most commonly, liquids and gases are sent, but pneumatic tubes using compressed air can also transport solid capsules.

As for gases and liquids, any chemically stable substance can be sent through a pipeline. Therefore sewage, slurry, water, or even beer pipelines exist; but arguably the most valuable are those transporting crude petroleum and refined petroleum product including fuels: oil (oleoduct), natural gas (gas grid), and biofuels.

Types by Transported Substance

For Oil or Natural Gas: It is uncertain when the first crude oil pipeline was built. Credit for the development of pipeline transport is disputed, with competing claims for Vladimir Shukhov and the Branobel company in the late 19th century, and the Oil Transport Association, which first constructed a 2-inch (51 mm) wrought iron pipeline over a 6-mile (9.7 km) track from an oil field in Pennsylvania to a railroad station in Oil Creek, in the 1860s. Pipelines are generally the most economical way to transport large quantities of oil, refined oil products or natural gas over land. Natural gas is methane and is found in nature mixed with other gases, mainly propane and butane. LPG as propane or LP gas and butane are wet gases when under light pressure are liquids and thus are called Natural Gas Liquids or NGL's.

These small NGL processing facilities can be located in oil fields so the butane and propane liquid under light pressure of 125 pounds per square inch, can be shipped by rail, truck or pipeline. Propane can be used as a fuel in oil fields to heat various facilities used by the oil drillers or equipment and trucks used in the oil patch. EG: Propane will convert from a gas to a liquid under light pressure under 40 PSI, give or take depending on temperature, and is pumped into cars and trucks at less than 125 PSI at retail stations. Pipelines and rail cars use about double that pressure to pump at 250 PSI. To convert PSI pressure to metric pascals, multiply one PSI is 6,894.75 pascals which is the same as 2.0360 inches of mercury.

The distance to ship propane to markets is much shorter as thousands of NGL processing plants are located in oil fields or close by when a number of pipelines tie into each other from various relatively close fields. Many Bakken Basin oil companies in North Dakota, Montana, Manitoba and Saskatchewan gas fields separate the NGL's in the field, allowing the drillers to sell propane directly to small wholesalers, eliminating the large refinery control of product and prices for propane or butane. One oil company operating in the Bakken in ND, stated two weeks ago, that the estimate of 577 billion barrels of oil-gas in the Bakken, should be upgraded to 903 billion barrels.

Present industry figures show about 60 percent of propane-butane was produced BEGO from gas fields while the other 40 percent was produced at oil refineries when a barrel of oil is cracked or refined into various products. With the huge amount of oil and gas fracking or fracing or fracture of oil bearing rock around North America over the past 6 years, i would estimate that now about 70 percent of propane-butane mixed with other NGL's, is produced from oil and gas fields, with the balance of 30 percent of propane-butane produced from oil refineries when oil is cracked. Both propane used as auto fuel HD5 grade and butane, bring a higher value to any drilling company, so they explore the WET gas fields first, to bring in more money than a dryer gas well.

The most recent major pipeline to start operating in North America, is a TransCanada natural gas line going north across the Niagara region bridges with Marcellus shale gas from Pennsylvania and others tied in methane or natural gas sources, into the Canadian province of Ontario as of the fall of 2012, supplying 16 percent of all the natural gas used in Ontario.

This new U.S. supplied natural gas displaces the natural gas formerly shipped to Ontario from western Canada in Alberta and Manitoba, thus dropping the government regulated pipeline shipping charges because of the significantly shorter distance from gas source to consumer. Compared to shipping by railroad, pipelines have lower cost per unit and higher capacity.

To avoid delays and U.S. government regulation, many small, medium and large oil producers in North Dakota, have decided to run an oil pipeline north to Canada to meet up with a Canadian oil pipeline shipping oil from west to east. This allows the Bakken Basin and Three Forks oil producers to get higher negotiated prices for their oil because they will not be restricted to just one wholesale market in the U.S. The distance from the biggest oil patch in ND, is Williston, ND, only about 85 miles or 137 kilometres to the Canadian border and Manitoba. Mutual funds and joint ventures are big investors in new oil and gas pipelines as they are very long term and have set fees that both U.S and Canada that guarantee a profit to pipe the oil or gas or NGL. Recently in the fall of 2012, the U.S. is now exporting propane to Europe, known as LPG, as wholesale prices there are much higher than in North America.

As more North American pipelines are built, even more exports of LNG, propane and butane etc., will occur on all three U.S. coasts. To give insight, North Dakota's oil production has grown to 5 times in late 2012 compared to what it was just 6 years ago creating thousands of good paying long term jobs. ND oil companies are shipping huge amounts of oil by tanker rail car as they can direct the oil to the market that gives the best price but pipelines are cheaper. Rail cars can be used to avoid a congested oil pipeline to get the oil to a different pipeline and get the oil to market faster and different less busy oil refineries.

Canada is reversing and converting the major natural gas pipeline going from east-to-west and expanding it and using it to ship western Canadian oil eastward. From a presently rated 250,000 barrels equivalent per day pipeline, it will be expanded to between one million to 1.3 million barrels per day to ship oil instead of natural gas. Since this natural gas pipeline already exists, it will bring western oil to refineries in Ontario, Michigan, Ohio, PA, Quebec and New York by early 2014. New Brunswick will also refine some of this western Canadian crude and export some crude and refined oil to Europe from its deep water oil ULCC loading port. This oil will displace Brent crude imported into any U.S. eastern seaboard refinery, wanting to buy much cheaper North American oil, than importing it from overseas and Central and South America. The normal price spread between imported light Brent oil and WTI, West Texas Intermediate oil, is about $15 to $20 per barrel and as high as $30 in 2012. Alaska and Louisiana both have very light crude and the spot price for it is very close to the daily Brent price.

As of 17 December 2012, the spot price for West Texas Intermediate (WTI) oil delivered (primarily via pipeline) at Cushing, Oklahoma was $86.71 while Brent oil (a benchmark price for oil delivered all over the world, primarily via ship) was $109.35. A spread of more than $22 per barrel.

Additional future pipelines should drop both the spot wholesale and long term contract prices of oil for eastern seaboard U.S and Canadian refineries. Refineries may be able to drop wholesale prices of gasoline, diesel fuel, home heating oil and jet fuel while making higher profits than they are making now once the western Canadian oil starts to flow east in the 2014 Q1 completed expanded oil pipeline.

Some parts of that pipeline will be pumping oil sometime in 2013 to Ontario crossing into Port Huron, Michigan from Sarnia, Ontario. That oil pipeline, which also carries North Dakota crude oil through Canada, can be expanded to triple or quadruple its planned one million barrel per day limit when required. This Canadian oil pipeline is a huge threat to OPEC and their control of world market prices of crude oil. Maybe we should call this new supply of western oil COP, or Canadian Oil Price.

Although pipelines can be built under the sea, that process is economically and technically demanding, so the majority of oil at sea is transported by tanker ships.

The market size for oil and gas pipeline construction experienced tremendous growth prior to the economic downturn in 2008. The industry grew from $23 billion in 2006 to $39 billion in 2008. After faltering in 2009, demand for pipeline expansion and updating increased the following year as energy production grew. By 2012, almost 32,000 miles of North American pipeline were being planned or under construction. Several major North American pipelines under development in 2012 included:

- Sand Hills Pipeline: This pipeline is being developed by DCP Midstream. It will extend for 700 miles from West to East Texas and have an initial capacity of approximately 120,000 per day. The first segment of the pipeline opened in October 2012.
- Pecos River Pipeline: Developed by Bridger Group and Advantage Pipeline, the Pecos River Pipeline will provide a link from the Delaware Basin to the Gulf Coast. The pipeline will run from the Pecos, Texas, to Crane, Texas, where it will connect to the Longhorn Pipeline. The Pecos River Pipeline will have an initial capacity of 150,000 barrels per day.
- Sterling III Pipeline: Under development from Oneok Partners, the Sterling III Pipeline will extend from parts of Texas and Oklahoma in the Mid-Continent region to the Gulf Coast, and have a capacity of approximately 193,000 barrels per day.
- ATEX Express: Enterprise Products Partners is developing the 1,230-mile ATEX Express pipeline to transport ethane from Pennsylvania, West Virginia, and Ohio to the Gulf Coast.

As of December 2012, construction was halted on TransCanada's Keystone XL pipeline in East Texas after a landowner filed a lawsuit claiming TransCanada misrepresented the type of fuel the pipeline would transport. The pipeline is awaiting a presidential permit after TransCanada was required to reroute the pipeline in portions of Nebraska to avoid environmentally sensitive areas.

Oil pipelines are made from steel or plastic tubes with inner diameter typically from 4 to 48 inches (100 to 1,200 mm). Most pipelines are typically buried at a depth of about 3 to 6 feet (0.91 to 1.8 m). To protect pipes from impact, abrasion, and corrosion, a variety of methods are used. These can include wood lagging (wood slats), concrete coating, rockshield, high-density polyethylene, imported sand padding, and padding machines.

The oil is kept in motion by pump stations along the pipeline, and usually flows at speed of about 1 to 6 metres per second (3.3 to 20 ft/s). Multi-product pipelines are used to transport two or more different products in sequence in the same pipeline. Usually in multi-product pipelines there is no physical separation between the different products. Some mixing of adjacent products occurs, producing interface, also known in the industry as "transmix." At the receiving facilities this interface is usually absorbed in one of the products based on pre-calculated absorption rates. Alternately, transmix may be diverted and shipped to facilities for separation of the commingled products.

Crude oil contains varying amounts of paraffin wax and in colder climates wax buildup may occur within a pipeline. Often these pipelines are inspected and cleaned using pipeline inspection gauges, *pigs*, also known as *scrapers* or *Go-devils*. Smart pigs (also known as intelligent or intelligence pigs) are used to detect anomalies in the pipe such as dents, metal loss caused by corrosion, cracking or other mechanical damage. These devices are launched from pig-launcher stations and travel through the pipeline to be received at any other station down-stream, either cleaning wax deposits and material that may have accumulated inside the line or inspecting and recording the condition of the line.

For natural gas, pipelines are constructed of carbon steel and vary in size from 2 to 60 inches (51 to 1,500 mm) in diameter, depending on the type of pipeline. The gas is pressurized by compressor stations and is odourless unless mixed with a mercaptan odorant where required by a regulating authority.

For Ammonia

Highly toxic ammonia is theoretically the most dangerous substance to be transported through long-distance pipelines. However, incidents on ammonia-transporting lines are uncommon - unlike on industrial ammonia-processing equipment.

A major example of ammonia pipeline is the Ukrainian *Transammiak* line connecting the TogliattiAzot facility in Russia to the exporting Black Sea-port of Odessa.

For Biofuels (Ethanol and Biobutanol)

Pipelines have been used for transportation of ethanol in Brazil, and there are several ethanol pipeline projects in Brazil and the United States. Main problems related to the shipment of ethanol by pipeline are its high oxygen content, which makes it corrosive, and absorption of water and impurities in pipelines, which is not a problem with oil and natural gas. Insufficient volumes and cost-effectiveness are other considerations limiting construction of ethanol pipelines.

For Coal and Ore

Slurry pipelines are sometimes used to transport coal or ore from mines. The material to be transported is closely mixed with water before being introduced to the pipeline; at the far end, the material must be dried. One example is a 525 km slurry pipeline which is planned to transport iron ore from the Minas-Rio mine (producing 26.5 million tonnes per year) to a port at Açu in Brazil. An existing example is the 85 km Savage River Slurry pipeline in Tasmania, Australia, possibly the world's first when it was built in 1967. It includes a 366m bridge span at 167m above the Savage River.

For Hydrogen

Hydrogen pipeline transport is a transportation of hydrogen through a pipe as part of the hydrogen infrastructure. Hydrogen pipeline transport is used to connect the point of hydrogen production or delivery of hydrogen with the point of demand, with transport costs similar to CNG, the technology is proven. Most hydrogen is produced at the place of demand with every 50 to 100 miles (160 km) an industrial production facility. The 1938 - Rhine-Ruhr 240 km hydrogen pipeline is still in operation. As of 2004 there are 900 miles (1,400 km) of low pressure hydrogen pipelines in the USA and 930 miles (1,500 km) in Europe.

For Water

Two millennia ago the ancient Romans made use of large aqueducts to transport water from higher elevations by building the aqueducts in graduated segments that allowed gravity to push the water along until it reached its destination. Hundreds of these were built throughout Europe and elsewhere, and along with flour mills were considered the lifeline of the Roman Empire. The ancient Chinese also made use of channels and pipe systems for public works. The famous Han Dynasty court eunuch Zhang Rang (d. 189 AD) once ordered the engineer Bi Lan to construct a series of square-pallet chain pumps outside the capital city of Luoyang. These chain pumps serviced the imperial palaces and living quarters of the capital city as the water lifted by the chain pumps was brought in by a stoneware pipe system.

Pipelines are useful for transporting water for drinking or irrigation over long distances when it needs to move over hills, or where canals or channels are poor choices due to considerations of evaporation, pollution, or environmental impact.

The 530 km (330 mi) Goldfields Water Supply Scheme in Western Australia using 750 mm (30 inch) pipe and completed in 1903 was the largest water supply scheme of its time.

Examples of significant water pipelines in South Australia are the Morgan-Whyalla (completed 1944) and Mannum-Adelaide (completed 1955) pipelines, both part of the larger Snowy Mountains scheme. There are two Los Angeles, California aqueducts, the *Owens Valley aqueduct* (completed 1913) and the *Second Los Angeles Aqueduct* (completed 1970) which also include extensive use of pipelines.

The Great Manmade River of Libya supplies 3,680,000 cubic metres (4,810,000 cu yd) of water each day to Tripoli, Benghazi, Sirte, and several other cities in Libya. The pipeline is over 2,800 kilometres (1,700 mi) long, and is connected to wells tapping an aquifer over 500 metres (1,600 ft) underground.

For Beer

Bars in the Veltins-Arena, a major football ground in Gelsenkirchen, Germany, are interconnected by a 5 km long beer pipeline. It is the favoured method for distributing beer in such large stadiums, because the bars have to overcome big differences between demands during various stages of a match; this allows them to be supplied by a central tank.

In Randers city in Denmark, the so-called Thor beer pipeline still exists. Originally copper pipes ran directly from the brewery and, when in the 90s the brewery moved out of the city, Thor beer replaced the centre of a star with a giant tank.

For Other Uses

The village of Hallstatt in Austria, which is known for its long history of salt mining, claims to contain “the oldest industrial pipeline in the world”, dating back to 1595. It was constructed from 13,000 hollowed out tree trunks to transport salt water for 40 kilometres from Hallstatt to Ebensee.

Types by Transport Function

In general, pipelines can be classified in three categories depending on purpose:

Gathering Pipelines: Group of smaller interconnected pipelines forming complex networks with the purpose of bringing crude oil or natural gas from several nearby wells to a treatment plant or processing facility. In this group, pipelines are usually short- a couple of hundred metres- and with small diameters. Also sub-sea pipelines for collecting product from deep water production platforms are considered gathering systems.

Transportation Pipelines: Mainly long pipes with large diameters, moving products (oil, gas, refined products) between cities, countries and even continents. These transportation networks include several compressor stations in gas lines or pump stations for crude and multiproducts pipelines.

Distribution Pipelines: Composed of several interconnected pipelines with small diameters, used to take the products to the final consumer. Feeder lines to distribute gas to homes and businesses downstream. Pipelines at terminals for distributing products to tanks and storage facilities are included in this group.

Construction

When a pipeline is built, the construction project not only covers the civil work to lay the pipeline and build the pump/compressor stations, it also has to cover all the work related to the installation of the field devices that will support remote operation.

The pipeline is routed along what is known as a "right of way". Pipelines are generally developed and built using the following stages:

1. Open season to determine market interest: Potential customers are given the chance to sign up for part of the new pipeline's capacity rights. This stage lasts up to two months. If interest in the pipeline is limited, the project does not move forward.
2. Route (right of way) selection
3. Pipeline design: The pipeline project may take a number of forms, including the construction of a new pipeline, conversion of existing pipeline from one fuel type to another, or improvements to facilities on a current pipeline route. The design process may take up to six months.
4. Obtaining approval: Once the design is finalized and the first pipeline customers have purchased their share of capacity, the

project must be approved by the relevant regulatory agencies. The process can last up to 18 months, particularly for pipelines that span multiple states.

5. Surveying the route
6. Clearing the route
7. Trenching - Main Route and Crossings (roads, rail, other pipes, etc.)
8. Installing the pipe
9. Installing valves, intersections, etc.
10. Covering the pipe and trench
11. Testing: Once construction is completed, the new pipeline is subjected to tests to ensure its structural integrity. These may include hydrostatic testing and line packing.

Russia has "Pipeline Troops" as part of the Rear Services, who are trained to build and repair pipelines. Russia is the only country to have Pipeline Troops.

Operation

Field devices are instrumentation, data gathering units and communication systems. The field Instrumentation includes flow, pressure and temperature gauges/transmitters, and other devices to measure the relevant data required. These instruments are installed along the pipeline on some specific locations, such as injection or delivery stations, pump stations (liquid pipelines) or compressor stations (gas pipelines), and block valve stations.

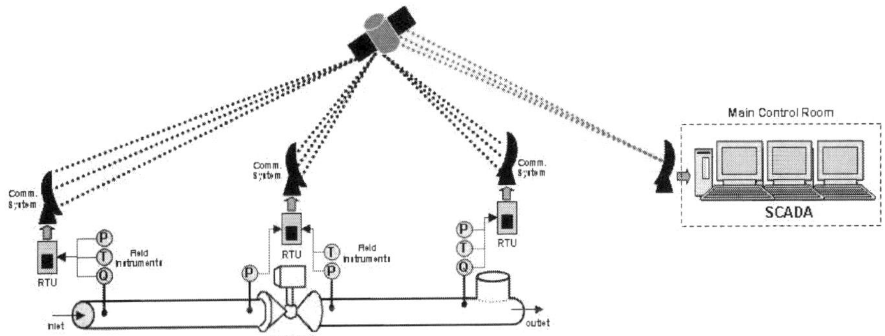

***Figure:** The SCADA System for pipelines.*

The information measured by these field instruments is then gathered in local Remote Terminal Units (RTU) that transfer the field data to a central location in real time using communication systems, such as satellite channels, microwave links, or cellular phone connections. Pipelines are controlled and operated remotely, from what is usually known as *The Main Control Room*. In this centre, all the data related to field measurement is consolidated in one central database. The data is received from multiple RTUs along the pipeline. It is common to find RTUs installed at every station along the pipeline.

The SCADA system at the Main Control Room receives all the field data and presents it to the pipeline operator through a set of screens or Human Machine Interface, showing the operational conditions of the pipeline. The operator can monitor the hydraulic conditions of the line, as well as send operational commands (open/close valves, turn on/off compressors or pumps, change setpoints, etc.) through the SCADA system to the field.

To optimize and secure the operation of these assets, some pipeline companies are using what is called *Advanced Pipeline Applications*, which are software tools installed on top of the SCADA system, that provide extended functionality to perform leak detection, leak location, batch tracking (liquid lines), pig tracking, composition tracking, predictive modelling, look ahead modelling, operator training and more.

Technology

Components: Pipeline networks are composed of several pieces of equipment that operate together to move products from location to location. The main elements of a pipeline system are:

Initial Injection Station: Known also as *supply* or *inlet* station, is the beginning of the system, where the product is injected into the line. Storage facilities, pumps or compressors are usually located at these locations.

Compressor/Pump Stations: Pumps for liquid pipelines and Compressors for gas pipelines, are located along the line to move the product through the pipeline. The location of these stations is defined by the topography of the terrain, the type of product being transported, or operational conditions of the network.

Partial Delivery Station: Known also as *intermediate* stations, these facilities allow the pipeline operator to deliver part of the product being transported.

Block Valve Station: These are the first line of protection for pipelines. With these valves the operator can isolate any segment of the line for maintenance work or isolate a rupture or leak. Block valve stations are usually located every 20 to 30 miles (48 km), depending on the type of pipeline. Even though it is not a design rule, it is a very usual practice in liquid pipelines. The location of these stations depends exclusively on the nature of the product being transported, the trajectory of the pipeline and/or the operational conditions of the line.

Regulator Station: This is a special type of valve station, where the operator can release some of the pressure from the line. Regulators are usually located at the downhill side of a peak.

Final Delivery Station: Known also as outlet stations or terminals, this is where the product will be distributed to the consumer. It could be a tank terminal for liquid pipelines or a connection to a distribution network for gas pipelines.

Leak Detection Systems

Since oil and gas pipelines are an important asset of the economic development of almost any country, it has been required either by government regulations or internal policies to ensure the safety of the assets, and the population and environment where these pipelines run.

Pipeline companies face government regulation, environmental constraints and social situations. Government regulations may define minimum staff to run the operation, operator training requirements; pipeline facilities, technology and applications required to ensure operational safety. For example, in the State of Washington it is mandatory for pipeline operators to be able to detect and locate leaks of 8 percent of maximum flow within fifteen minutes or less. Social factors also affect the operation of pipelines. In third world countries, product theft is a problem for pipeline companies. It is common to find unauthorized extractions in the middle of the pipeline. In this case, the detection levels should be under two percent of maximum flow, with a high expectation for location accuracy.

Various technologies and strategies have been implemented for monitoring pipelines, from physically walking the lines to satellite surveillance. The most common technology to protect pipelines from occasional leaks is Computational Pipeline Monitoring or CPM. CPM takes information from the field related to pressures, flows, and temperatures to estimate the hydraulic behaviour of the product being transported. Once the estimation is completed, the results are compared to other field references to detect the presence of an anomaly or unexpected situation, which may be related to a leak.

The American Petroleum Institute has published several articles related to the performance of CPM in liquids pipelines, the API Publications are:

- API 1130 – Computational pipeline monitoring for liquids pipelines
- API 1155 – Evaluation methodology for software based leak detection systems
- API 1149 – Pipeline variable uncertainties & their effects on leak detectability

Implementation

As a rule pipelines for all uses are laid in most cases underground. However in some cases it is necessary to cross a valley or a river on a pipeline bridge. Pipelines for centralized heating systems are often laid on the ground or overhead. Pipelines for petroleum running through permafrost areas as Trans-Alaska-Pipeline are often run overhead in order to avoid melting the frozen ground by hot petroleum which would result in sinking the pipeline in the ground.

Maintenance

Maintenance of pipelines includes checking Cathodic protection levels for the proper range, surveillance for construction, erosion, or leaks by foot, land vehicle, boat, or air, and running cleaning pigs, when there is anything carried in the pipeline that is corrosive.

US pipeline maintenance rules are covered in Code of Federal Regulations(CFR) sections, 49 CFR 192 for natural gas pipelines, and 49 CFR 195 for petroleum liquid pipelines.

Regulation

In the US, onshore and offshore pipelines used to transport oil and gas are regulated by the Pipeline and Hazardous Materials Safety Administration (PHMSA). Certain offshore pipelines used to produce oil and gas are regulated by the Minerals Management Service (MMS). In Canada, pipelines are regulated by either the provincial regulators or, if they cross provincial boundaries or the Canada/US border, by the National Energy Board (NEB). Government regulations in Canada and the United States require that buried fuel pipelines must be protected from corrosion. Often, the most economical method of corrosion control is by use of pipeline coating in conjunction with cathodic protection and technology to monitor the pipeline. Above ground, cathodic protection is not an option. The coating is the only external protection.

Pipelines and Geopolitics

Pipelines for major energy resources (petroleum and natural gas) are not merely an element of trade. They connect to issues of geopolitics and international security as well, and the construction, placement, and control of oil and gas pipelines often figure prominently in state interests and actions. A notable example of pipeline politics occurred at the beginning of the year 2009, wherein a dispute between Russia and Ukraine ostensibly over pricing led to a major political crisis. Russian state-owned gas company Gazprom cut off natural gas supplies to Ukraine after talks between it and the Ukrainian government fell through. In addition to cutting off supplies to Ukraine, Russian gas flowing through Ukraine—which included nearly all supplies to Southeastern Europe and some supplies to Central and Western Europe—was cut off, creating a major crisis in several countries heavily dependent on Russian gas as fuel. Russia was accused of using the dispute as leverage in its attempt to keep other powers, and particularly the European Union, from interfering in its "near abroad".

Oil and gas pipelines also figure prominently in the politics of Central Asia and the Caucasus.

Hazard Identification

Because the solvent fraction of dilbit typically comprises volatile aromatics like naptha and benzene, reasonably rapid carrier vapourisation can be expected to follow an above-ground spill—ostensibly enabling timely intervention by leaving only a viscous residue that is slow to migrate. Effective protocols to minimize exposure to petrochemical vapours are well-established, and oil spilled from the pipeline would be unlikely to reach the aquifer unless incomplete remediation were followed by the introduction of another carrier (e.g., a series of torrential downpours).

The Keystone XL extension is designed to be buried under four feet of soil, which will hinder post-spill vapourisation of the carrier fraction. Diluent and bitumen will migrate at different rates, depending on the temperature- and composition of the surrounding soils, but separation will take place more slowly as the aromatics diffuse through sediment rather than through air.

The introduction of benzene and other volatile organic compounds (collectively BTEX) to the subterranean environment compounds the threat posed by a pipeline leak. Particularly if followed by rain, a pipeline breach would result in BTEX dissolution and equilibration of benzene in water, followed by percolation of the admixture into the aquifer. Benzene can cause many health problems and is carcinogenic with EPA Maximum Contaminant Level (MCL) set at 5 µg/L for potable water. Although it is not well studied, single benzene exposure events have been linked to acute carcinogenesis. Additionally, the exposure of livestock, mainly cattle, to benzene has been shown to cause many health issues, such as neurotoxicity, fetal damage and fatal poisoning.

The entire surface of an above-ground pipeline can be directly examined for material breach. Pooled petroleum is unambiguous, readily spotted, and indicates the location of required repairs. Because the effectiveness of remote inspection is limited by the cost of monitoring equipment, gaps between sensors, and data that requires interpretation, leaks in buried pipe are more likely to go undetected

Pipeline developers do not always prioritise effective surveillance against leaks. Buried pipes draw fewer complaints. They are insulated from extremes in ambient temperature, they are shielded from ultraviolet rays, and they are less exposed to photodegradation. Buried pipes are isolated from airborne debris, electrical storms, tornadoes, hurricanes, hail, and acid rain. They are protected from nesting birds, rutting mammals, and wayward buckshot. Buried pipe is less vulnerable to accident damage (e.g., automobile collisions) and less accessible to vandals, saboteurs, and terrorists.

Exposure

Previous work has shown that a 'worst-case exposure scenario' can be limited to a specific set of conditions. Based on the advanced detection methods and pipeline shut-off SOP developed by TransCanada, the risk of a substantive or large release over a short period of time contaminating groundwater with benzene is unlikely. Detection, shutoff, and remediation procedures would limit the dissolution and transport of benzene. Therefore the exposure of benzene would be limited to leaks that are below the limit of detection and go unnoticed for extended periods of time. Leak detection is monitored through a SCADA system that assesses pressure and volume flow every 5 seconds. A pinhole leak that releases small quantities that cannot be detected by the SCADA system (<1.5% flow) could accumulate into a substantive spill. Detection of pinhole leaks would come from a visual or olfactory inspection, aerial surveying, or mass-balance inconsistencies. It is assumed that pinhole leaks are discovered within the 14 day inspection interval, however snow cover and location (e.g. remote, deep) could delay detection. Benzene typically makes up

0.1 – 1.0% of oil and will have varying degrees of volatility and dissolution based on environmental factors.

Even with pipeline leak volumes within SCADA detection limits, sometimes pipeline leaks are misinterpreted by pipeline operators to be pump malfunctions, or other problems. The Enbridge Line 6B crude oil pipeline failure in Marshall, Michigan on July 25, 2010 was thought by operators in Edmonton to be from column separation of the dilbit in that pipeline. The leak in wetlands along the Kalamazoo River was only confirmed 17 hours after it happened by a local gas company employee in Michigan.

Spill Frequency-Volume

Although the Pipeline and Hazardous Materials Safety Administration (PHMSA) has standard baseline incident frequencies to estimate the number of spills, TransCanada altered these assumptions based on improved pipeline design, operation, and safety. Whether these adjustments are justified is debatable as these assumptions resulted in a nearly 10-fold decrease in spill estimates. Given that the pipeline crosses 247 miles of the Ogallala Aquifer, or 14.5% of the entire pipeline length, and the 50-year life of the entire pipeline is expected to have between 11 – 91 spills, approximately 1.6 – 13.2 spills can be expected to occur over the aquifer. An estimate of 13.2 spills over the aquifer, each lasting 14 days, results in 184 days of potential exposure over the 50 year lifetime of the pipeline. In the reduced scope 'worst case exposure scenario,' the volume of a pinhole leak at 1.5% of max flow-rate for 14 days has been estimated at 189,000 barrels or 7.9 million gallons of oil. According to PHMSA's incident database, only 0.5% of all spills in the last 10 years were >10,000 barrels.

Benzene Fate and Transport

Benzene is considered a light aromatic hydrocarbon with high solubility and high volatility. It is unclear how temperature and depth would impact the volatility of benzene, so assumptions have been made that benzene in oil (1% weight by volume) would not volatilize before equilibrating with water. Using the octanol-water partition coefficient and a 100-year precipitation event for the area, a worst-case estimate of 75 mg/L of benzene is anticipated to flow toward the aquifer. The actual movement of the plume through groundwater systems is not well described, although one estimate is that up to 4.9 billion gallons of water in the Ogallala Aquifer could become contaminated with benzene at concentrations above the MCL. The

Final Environmental Impact Statement from the State Department does not include a quantitative analysis because it assumed that most benzene will volatilize.

Previous Dilbit Spill Remediation Difficulties

One of the major concerns about dilbit is the difficulty in cleaning it up. Enbridge's Line 6B, a 30 inch crude oil pipeline, ruptured in Marshall, Michigan on July 25, 2010, mentioned above, spilled at least 843,000 gallons of dilbit. After detection of the leak, booms and vacuum trucks were deployed. Heavy rains caused the river to overtop existing dams, and carried dilbit 30 miles downstream before the spill was contained. Remediation work collected over 1.1 million gallons of oil and almost 200,000 cubic yards of oil-contaminated sediment and debris from the Kalamazoo River system. However, oil was still being found in affected waters in October 2012.

Dangers

Accidents: Pipelines conveying flammable or explosive material, such as natural gas or oil, pose special safety concerns.

- June 16, 1976 - A gasoline pipeline was ruptured by a road construction crew in Los Angeles, California. Gasoline sprayed across the area, and soon ignited, killing 9, and injuring at least 14 others. Confusion over the depth of the pipeline in the construction area seemed to be a factor in the accident.
- 1982 - One of the largest non-nuclear explosions in history occurred along the Trans-Siberian Pipeline in the former Soviet Union. It has been alleged that the explosion was the result of CIA sabotage of the Trans-Siberian Pipeline.
- June 4, 1989 - sparks from two passing trains detonated gas leaking from an LPG pipeline near Ufa, Russia. Up to 645 people were reported killed.
- October 17, 1998 - 1998 Jesse pipeline explosion at Jesse in the Niger Delta in Nigeria, a petroleum pipeline exploded killing about 1,200 villagers, some of whom were scavenging gasoline - the worst of several similar incidents in this country.
- June 10, 1999 - a pipeline rupture in a Bellingham, Washington park led to the release of 277,200 gallons of gasoline. The gasoline was ignited, causing an explosion that killed two children and one adult. Misoperation of the pipeline and a previously damaged section of the pipe that was not detected before were identified as causing the failure.

- August 19, 2000 - natural gas pipeline rupture and fire near Carlsbad, New Mexico; this explosion and fire killed 12 members of an extended family. The cause was due to severe internal corrosion of the pipeline.
- July 30, 2004 - a major natural gas pipeline exploded in Ghislenghien, Belgium near Ath (thirty kilometres southwest of Brussels), killing at least 24 people and leaving 132 wounded, some critically. (Expatica)
- May 12, 2006 - an oil pipeline ruptured outside Lagos, Nigeria. Up to 200 people may have been killed.
- November 1, 2007 - a propane pipeline exploded near Carmichael, Mississippi, about 30 miles (48 km) south of Meridian, Mississippi. Two people were killed instantly and an additional four were injured. Several homes were destroyed and sixty families were displaced. The pipeline is owned by Enterprise Products Partners LP, and runs from Mont Belvieu, Texas, to Apex, North Carolina. Inability to find flaws in pre-1971 ERW seam welded pipe flaws was a contributing factor to the accident.
- September 9, 2010 - a 30 inch diameter high pressure natural gas pipeline owned by Pacific Gas & Electric exploded in flames in the Crestmoor residential neighbourhood 2 mi (3.2 km) west of San Francisco International Airport, killing 8, injuring 58, and destroying 38 homes. Poor quality control of the pipe used & of the construction were cited as factors in the accident.

As Targets

Pipelines can be the target of vandalism, sabotage, or even terrorist attacks. In war, pipelines are often the target of military attacks, as destruction of pipelines can seriously disrupt enemy logistics.

Marine Pipelines

In places, a pipeline may have to cross water expanses, such as small seas, straights and rivers. In many instances, they lie entirely on a seabed. These pipelines are referred to as marine pipelines (also: submarine or offshore). They are used primarily to carry oil or gas, but transportation of water is also important. In offshore projects, a distinction is made between a *flowline* and a pipeline. The former is an *intrafield* pipeline, in the sense that it is used to connect subsea wellheads, manifolds and the platform *within* a particular development field. The latter, sometimes referred to as an *export pipeline*, is used to bring the resource to shore. The construction and maintenance of marine pipelines imply logistical challenges that are different from those onland, mainly because of wave and current dynamics, along with other geohazards.

Hydraulic Structure

A hydraulic structure is structure submerged or partially submerged in any body of water, which disrupts the natural flow of water. They can be used to divert, disrupt or completely stop the flow. An example of a hydraulic structure would be a dam, which slows the normal flow rate of the river in order to power turbines. A hydraulic structure can be built in rivers, a sea, or any body of water where there is a need for a change in the natural flow of water.

Hydraulic structures may also be used to measure the flow of water. When used to measure the flow of water, hydraulic structures are defined as a class of specially shaped, static devices over or through which water is directed in such a way that under free-flow conditions at a specified location (point of measurement) a known level to flow relationship exists. Hydraulic structures of this type can generally be divided into two categories: flumes and weirs.

River Engineering

River engineering is the process of planned human intervention in the course, characteristics, or flow of a river with the intention of producing some defined benefit. People have intervened in the natural course and behaviour of rivers since before recorded history - to manage the water resources, to protect against flooding, or to make passage along or across rivers easier. From Roman times, rivers have been used as a source of hydropower. From the late 20th century, river engineering has had environmental concerns broader than immediate human benefit and some river engineering projects have been concerned exclusively with the restoration or protection of natural characteristics and habitats.

Hydromodification is a term which encompasses the systematic response to alterations to riverine and non-riverine water bodies such as coastal waters (estuaries and bays) and lakes. The U.S. Environmental Protection Agency (EPA) has defined hydromodification as the "alteration of the hydrologic characteristics of coastal and non-coastal waters, which in turn could cause degradation of water resources." River engineering has often resulted in unintended systematic responses.

The river engineering discipline now strives to repair hydromodified degradations and account for potential systematic response to planned alterations by considering fluvial geomorphology. Fluvial geomorphology is the study of how rivers change their form over time. Fluvial geomorphology is the cumulation of a number of sciences including open channel hydraulics, sediment transport, hydrology, physical geology, and riparian ecology. River engineering attempts to understand fluvial geomorphology, implement a physical alteration, and maintain public safety.

Characteristics of Rivers

The size of rivers above any tidal limit and their average freshwater discharge are proportionate to the extent of their basins and the amount of rain which, after falling over these basins, reaches the river channels in the bottom of the valleys, by which it is conveyed to the sea.

The basin of a river is the expanse of country bounded by a watershed (called a "divide" in North America) over which rainfall flows down towards the river traversing the lowest part of the valley, whereas the rain falling on the far slope of the watershed flows away to another river draining an adjacent basin. River basins vary in extent according to the configuration of the country, ranging from the insignificant drainage areas of streams rising on high ground very near the coast and flowing straight down into the sea, up to immense tracts of great continents, where rivers rising on the slopes of mountain ranges far inland have to traverse vast stretches of valleys and plains before reaching the ocean. The size of the largest river basin of any country depends on the extent of the continent in which it is situated, its position in relation to the hilly regions in which rivers generally arise and the sea into which they flow, and the distance between the source and the outlet into the sea of the river draining it.

The rate of flow of rivers depends mainly upon their fall, also known as the gradient or slope. When two rivers of different sizes have the same fall, the larger river has the quicker flow, as its retardation by friction against its bed and banks is less in proportion to its volume than is the case with the smaller river. The fall available in a section of a river approximately corresponds to the slope of the country it traverses; as rivers rise close to the highest part of their basins, generally in hilly regions, their fall is rapid near their source and gradually diminishes, with occasional irregularities, until, in traversing plains along the latter part of their course, their fall usually becomes quite gentle. Accordingly, in large basins, rivers in most cases begin as torrents with a very variable flow, and end as gently flowing rivers with a comparatively regular discharge.

The irregular flow of rivers throughout their course forms one of the main difficulties in devising works for mitigating inundations or for increasing the navigable capabilities of rivers. In tropical countries subject to periodical rains, the rivers are in flood during the rainy season and have hardly any flow during the rest of the year, while in temperate regions, where the rainfall is more evenly distributed throughout the year, evaporation causes the available rainfall to be much less in hot summer weather than in the winter months, so that the rivers fall to their low stage in the summer and are very liable to be in flood in the winter.

In fact, with a temperate climate, the year may be divided into a warm and a cold season, extending from May to October and from November to April in the Northern hemisphere respectively; the rivers are low and moderate floods are of rare occurrence during the warm period, and the rivers are high and subject to occasional heavy floods after a considerable rainfall during the cold period in most years. The only exceptions are rivers which have their sources amongst mountains clad with perpetual snow and are fed by glaciers; their floods occur in the summer from the melting of snow and ice, as exemplified by the Rhône above the Lake of Geneva, and the Arve which joins it below. But even these rivers are liable to have their flow modified by the influx of tributaries subject to different conditions, so that the Rhone below Lyon has a more uniform discharge than most rivers, as the summer floods of the Arve are counteracted to a great extent by the low stage of the Saône flowing into the Rhone at Lyon, which has its floods in the winter when the Arve, on the contrary, is low.

Another serious obstacle encountered in river engineering consists in the large quantity of detritus they bring down in flood-time, derived mainly from the disintegration of the surface layers of the hills and slopes in the upper parts of the valleys by glaciers, frost and rain. The power of a current to transport materials varies with its velocity, so that torrents with a rapid fall near the sources of rivers can carry down rocks, boulders and large stones, which are by degrees ground by attrition in their onward course into shingle, gravel, sand and silt, simultaneously with the gradual reduction in fall, and, consequently, in the transporting force of the current. Accordingly, under ordinary conditions, most of the materials brought down from the high lands by torrential water courses are carried forward by the main river to the sea, or partially strewn over flat alluvial plains during floods; the size of the materials forming the bed of the river or borne along by the stream is gradually reduced on proceeding seawards, so that in the Po River in Italy, for instance, pebbles and gravel are found for about 140 miles below Turin, sand along the next 100 miles, and silt and mud in the last 110 miles.

Methods

Improvements can be divided into those that are aimed at improving the flow of the river, particularly in flood conditions, and those that aim to hold back the flow, primarily for navigation purposes, although power generation is often an important factor. The former is known in the US as *channelization* and the latter is generally referred to as *canalization*.

Channelization

Reducing the length of the channel by substituting straight cuts for a winding course is the only way in which the (effective) fall can be increased. This involves some loss of capacity in the channel as a whole, and in the case of a large river with a considerable flow it is very difficult to maintain a straight cut owing to the tendency of the current to erode the banks and form again a sinuous channel. Even if the cut is preserved by protecting the banks, it is liable to produce changes shoals and raise the flood-level in the channel just below its termination. Nevertheless, where the available fall is exceptionally small, as in land originally reclaimed from the sea, such as the English Fenlands, and where, in consequence, the drainage is in a great measure artificial, straight channels have been formed for the rivers.

Because of the perceived value in protecting these fertile, low-lying lands from inundation, additional straight channels have also been provided for the discharge of rainfall, known as drains in the fens. Even extensive modification of the course of a river combined with an enlargement of its channel often produces only a limited reduction in flood damage. Consequently, such floodworks are only commensurate with the expenditure involved where significant assets (such as a town) are under threat. Additionally, even when successful, such floodworks may simply move the problem further downstream and threaten some other town. Recent floodworks in Europe have included restoration of natural floodplains and winding courses, so that floodwater is held back and released more slowly.

The removal of obstructions, natural or artificial (e.g., trunks of trees, boulders and accumulations of gravel) from a river bed furnishes a simple and efficient means of increasing the discharging capacity of its channel. Such removals will consequently lower the height of floods upstream. Every impediment to the flow, in proportion to its extent, raises the level of the river above it so as to produce the additional artificial fall necessary to convey the flow through the restricted channel, thereby reducing the total available fall.

Human intervention sometimes inadvertently modifies the course or characteristics of a river, for example by introducing obstructions such as mining refuse, sluice gates for mills, fish-traps, unduly wide piers for bridges and solid weirs. By impeding flow these measures can raise the flood-level upstream. Regulations for the management of rivers may include stringent prohibitions with regard to pollution, requirements for enlarging sluice-ways and the compulsory raising of their gates for the passage of floods, the removal of fish traps, which are frequently blocked up by leaves and floating rubbish, reduction in the number and width of bridge piers when rebuilt, and the substitution of movable weirs for solid weirs.

By installing gauges in a fairly large river and its tributaries at suitable points, and keeping continuous records for some time of the heights of the water at the various stations, the rise of the floods in the different tributaries, the periods they take in passing down to definite stations on the main river, and the influence they severally exercise on the height of the floods at these places, can be ascertained.

With the help of these records, and by observing the times and heights of the maximum rise of a particular flood at the stations on the various tributaries, the time of arrival and height of the top of the flood at any station on the main river can be predicted with remarkable accuracy two or more days beforehand. By communicating these particulars about a high flood to places on the lower river, weir-keepers are enabled to fully open the movable weirs beforehand to permit the passage of the flood, and riparian inhabitants receive timely warning of the impending inundation.

Where portions of a riverside town are situated below the maximum flood-level, or when it is important to protect land adjoining a river from inundations, the overflow of the river must be diverted into a flood-dam or confined within continuous embankments on both sides. By placing these embankments somewhat back from the margin of the river-bed, a wide flood-channel is provided for the discharge of the river as soon as it overflows its banks, while leaving the natural channel unaltered for the ordinary flow. Low embankments may be sufficient where only exceptional summer floods have to be excluded from meadows.

Occasionally the embankments are raised high enough to retain the floods during most years, while provision is made for the escape of the rare, exceptionally high floods at special places in the embankments, where the scour of the issuing current is guarded against, and the inundation of the neighbouring land is least injurious. In this manner, the increased cost of embankments raised above the highest flood-level of rare occurrence is avoided, as is the danger of breaches in the banks from an unusually high flood-rise and rapid flow, with their disastrous effects.

Effects

A most serious objection to the formation of continuous, high embankments along rivers bringing down considerable quantities of detritus, especially near a place where their fall has been abruptly reduced by descending from mountain slopes onto alluvial plains, is the danger of their bed being raised by deposit, producing a rise in the flood-level, and necessitating a raising of the embankments if inundations are to be prevented. Longitudinal sections of the Po River, taken in 1874 and 1901, show that its bed was materially raised during this period from the confluence of the Ticino to below Caranella, despite the clearance of sediment effected by the rush through breaches. Therefore, the completion of the embankments, together with their raising, would only eventually aggravate the injuries of the inundations they have been designed to prevent, as the escape of floods from the raised river must occur sooner or later.

In the UK, problems of flooding of domestic properties around the turn of the 21st century have been blamed on inadequate planning controls which have permitted development on floodplains. This exposes the properties on the floodplain to flood, and the substitution of concrete for natural strata speeds the run-off of water, which increases the danger of flooding downstream. In the Midwestern United States and the Southern United States the term for this measure is *channelization.* Much of it was done under the auspices or overall direction of the United States Army Corps of Engineers. One of the most heavily channelized areas in the United States is West Tennessee, where every major stream with one exception (the Hatchie River) has been partially or completely channelized.

Advantages

Channelization of a stream may be undertaken for several reasons. One is to make a stream more suitable for navigation or for navigation by larger vessels with deep draughts. Another is to restrict water to a certain area of a stream's natural bottom lands so that the bulk of such lands can be made available for agriculture. A third reason is flood control, with the idea of giving a stream a sufficiently large and deep channel so that flooding beyond those limits will be minimal or nonexistent, at least on a routine basis. One major reason is to reduce natural erosion; as a natural waterway curves back and forth, it usually deposits sand and gravel on the inside of the corners where the water flows slowly, and cuts sand, gravel, subsoil, and precious topsoil from the outside corners where it flows rapidly due to a change in direction. Unlike sand and gravel, the topsoil that is eroded does not get deposited on the inside of the next corner of the river. It simply washes away.

Disadvantages

Channelization has several predictable and negative effects. One of them is loss of wetlands. Wetlands are an excellent habitat for many forms of wildlife, and additionally serve as a "filter" for much of the world's surface fresh water. Another is the fact that channelized streams are almost invariably straightened. For example, the channelization of Florida's Kissimmee River has been cited as a cause contributing to the loss of wetlands. This straightening causes the streams to flow more rapidly, which can, in some instances, vastly increase soil erosion. It can also increase flooding downstream from the channelized area, as larger volumes of water travelling more rapidly than normal can reach choke points over a shorter period of time than they otherwise would, with a net effect of flood control in one area coming at the expense of greatly aggravated flooding in another.

In addition, studies have shown that stream channelization results in declines of river fish populations. A 1971 study of the Chariton River in northern Missouri, United States, found that the channelized section of the river contained only 13 species of fish, whereas the natural segment of the stream was home to 21 species of fish. The biomass of fish able to be caught in the dredged segments of the river was 80 percent less than in the natural parts of the same stream. This loss of fish diversity and abundance is thought to occur because of reduction in habitat, elimination of riffles and pools, greater fluctuation of stream levels and water temperature, and shifting substrates. The rate of recovery for a stream once it has been dredged is extremely slow, with many streams showing no significant recovery 30 to 40 years after the date of channelization.

Modern Policy

For the reasons cited above, in recent years stream channelization has been greatly curtailed in the U.S., and in some instances even partially reversed. The United States Government now has in place a "no net loss of wetlands" policy that means that stream channelization in one place has to be offset by the creation of new wetlands in another, a process known as "mitigation." The major agency involved in the enforcement of this policy is the same Army Corps of Engineers which for so long was the primary promoter of wide-scale channelization.

Often, in the instances where channelization is permitted, boulders may be installed in the bed of the new channel so that water velocity is slowed, and channels may be deliberately curved as well. In 1990 the U.S. Congress gave the Army Corps a specific mandate to include environmental protection in its mission, and in 1996 it authorized the Corps to undertake restoration projects. The U.S. Clean Water Act regulates certain aspects of channelization by requiring non-Federal entities (i.e. state and local governments, private parties) to obtain permits for dredging and filling operations. Permits are issued by the Army Corps with EPA participation.

This new policy also has its critics, however. Farmers who are losing land to soil erosion as channelized streams cease to be maintained feel particularly aggrieved. Not only are they losing their valuable property to erosion, the erosion ends up in the stream or river, and contributes to decreased water quality. They also point out that if such policies had been in place in the U.S. at the time of white settlement, the country probably would never have become a leader in world agriculture and a net exporter of food. Channelization critics respond that this is immaterial, as we are no longer living in the era of initial settlement.

Canalization of Rivers

Rivers whose discharge is liable to become quite small at their low stage, or which have a somewhat large fall, as is usual in the upper part of rivers, cannot be given an adequate depth for navigation purely by works which regulate the flow; their ordinary summer level has to be raised by impounding the flow with weirs at intervals across the channel, while a lock has to be provided alongside the weir, or in a side channel, to provide for the passage of vessels. A river is thereby converted into a succession of fairly level reaches rising in steps up-stream, providing still-water navigation comparable to a canal; but it differs from a canal in the introduction of weirs for keeping up the water-level, in the provision for the regular discharge of the river at the weirs, and in the two sills of the locks being laid at the same level instead of the upper sill being raised above the lower one to the extent of the rise at the lock, as usual on canals.

Canalization secures a definite available depth for navigation; and the discharge of the river generally is amply sufficient for maintaining the impounded water level, as well as providing the necessary water for locking. Navigation, however, is liable to be stopped during the descent of high floods, which in many cases rise above the locks; and it is necessarily arrested in cold climates on all rivers by long, severe frosts, and especially by ice. Many small rivers, like the Thames above its tidal limit, have been rendered navigable by canalization, and several fairly large rivers have thereby provided a good depth for vessels for considerable distances inland. Thus the canalized Seine has secured a navigable depth of $10^1/_2$ feet from its tidal limit up to Paris, a distance of 135 miles, and a depth of $6^3/_4$ feet up to Montereau, 62 miles higher up.

Regulation Works (Flow and Depth Control)

As rivers flow onward towards the sea, they experience a considerable diminution in their fall, and a progressive increase in the basin which they drain, owing to the successive influx of their various tributaries. Thus, their current gradually becomes more gentle and their discharge larger in volume and less subject to abrupt variations; and, consequently, they become more suitable for navigation. Eventually, large rivers, under favourable conditions, often furnish important natural highways for inland navigation in the lower portion of their course, as, for instance, the Rhine, the Danube and the Mississippi. River engineering works are only required to prevent changes in the course of the stream, to regulate its depth, and especially to fix the low-water channel and concentrate the flow in it, so as to increase as far as practicable the navigable depth at the lowest stage of the water level.

Engineering works to increase the navigability of rivers can only be advantageously undertaken in large rivers with a moderate fall and a fair discharge at their lowest stage, for with a large fall the current presents a great impediment to up-stream navigation, and there are generally great variations in water level, and when the discharge becomes very small in the dry season. It is impossible to maintain a sufficient depth of water in the low-water channel.

The possibility to secure uniformity of depth in a river by lowering the shoals obstructing the channel depends on the nature of the shoals. A soft shoal in the bed of a river is due to deposit from a diminution in velocity of flow, produced by a reduction in fall and by a widening of the channel, or to a loss in concentration of the scour of the main current in passing over from one concave bank to the next on the opposite side.

The lowering of such a shoal by dredging merely effects a temporary deepening, for it soon forms again from the causes which produced it. The removal, moreover, of the rocky obstructions at rapids, though increasing the depth and equalizing the flow at these places, produces a lowering of the river above the rapids by facilitating the efflux, which may result in the appearance of fresh shoals at the low stage of the river. Where, however, narrow rocky reefs or other hard shoals stretch across the bottom of a river and present obstacles to the erosion by the current of the soft materials forming the bed of the river above and below, their removal may result in permanent improvement by enabling the river to deepen its bed by natural scour.

The capability of a river to provide a waterway for navigation during the summer or throughout the dry season depends on the depth that can be secured in the channel at the lowest stage. The problem in the dry season is the small discharge and deficiency in scour during this period. A typical solution is to restrict the width of the low-water channel, concentrate all of the flow in it, and also to fix its position so that it is scoured out every year by the floods which follow the deepest part of the bed along the line of the strongest current. This can be effected by closing subsidiary low-water channels with dikes across them, and narrowing the channel at the low stage by low-dipping cross dikes extending from the river banks down the slope and pointing slightly up-stream so as to direct the water flowing over them into a central channel.

Estuarine Works

The needs of navigation may also require that a stable, continuous, navigable channel is prolonged from the navigable river to deep water at the mouth of the estuary. The interaction of river flow and tide needs to be modelled by computer or using scale models, moulded to the configuration of the estuary under consideration and reproducing in miniature the tidal ebb and flow and fresh-water discharge over a bed of very fine sand, in which various lines of training walls can be successively inserted. The models should be capable of furnishing valuable indications of the respective effects and comparative merits of the different schemes proposed for works.

Turbulence Modelling

Turbulence modelling is the construction and use of a model to predict the effects of turbulence. Averaging is often used to simplify the solution of the governing equations of turbulence, but models are needed to represent scales of the flow that are not resolved.

Closure Problem

A closure problem arises in the Reynolds-averaged Navier-Stokes (RANS) equation because of the non-linear term $\overline{\upsilon_i'\upsilon_j'}$ from the convective acceleration, known as the Reynolds stress,

$$R_{ij} = \overline{\upsilon_i'\upsilon_j'}$$

Closing the RANS equation requires modelling the Reynold's stress R_{ij}.

Eddy Viscosity

Joseph Boussinesq was the first practitioner of this (i.e. modelling the Reynold's stress), introducing the concept of eddy viscosity. In 1887 Boussinesq proposed relating the turbulence stresses to the mean flow to close the system of equations. Here the Boussinesq hypothesis is applied to model the Reynolds stress term. Note that a new proportionality constant $\nu_t > 0$, the turbulence eddy viscosity, has been introduced. Models of this type are known as eddy viscosity models or EVM's.

$$-\overline{\upsilon_i'\upsilon_j'} = \nu_t\left(\frac{\partial\overline{\upsilon}_i}{\partial x_j} + \frac{\partial\overline{\upsilon}_j}{\partial x_i}\right) - \frac{2}{3}\left(K + \nu_t\frac{\partial\overline{\upsilon}_k}{\partial x_k}\right)\delta_{ij}$$

Which can be written in shorthand as

$$-\overline{\upsilon_i'\upsilon_j'} = 2\nu_t S_{ij} - \frac{2}{3}K\delta_{ij}$$

where S_{ij} is the mean rate of strain tensor

ν_t is the turbulence eddy viscosity

$K = \frac{1}{2}\overline{\upsilon_i'^2}$ is the turbulence kinetic energy

and δ_{ij} is the Kronecker delta.

In this model, the additional turbulence stresses are given by augmenting the molecular viscosity with an eddy viscosity. This can be a simple constant eddy viscosity (which works well for some free shear flows such as axisymmetric jets, 2-D jets, and mixing layers).

Prandtl's Mixing-Length Concept

Later, Ludwig Prandtl introduced the additional concept of the mixing length, along with the idea of a boundary layer. For wall-bounded turbulent flows, the eddy viscosity must vary with distance from the wall, hence the addition of the concept of a 'mixing length'. In the simplest wall-bounded flow model, the eddy viscosity is given by the equation:

$$\nu_t = \left|\frac{\partial u}{\partial y}\right| l_m^2$$

where:

$\frac{\partial u}{\partial y}$ is the partial derivative of the streamwise velocity (u) with respect to the wall normal direction (y);

l_m is the mixing length.

This simple model is the basis for the "law of the wall", which is a surprisingly accurate model for wall-bounded, attached (not separated) flow fields with small pressure gradients.

More general turbulence models have evolved over time, with most modern turbulence models given by field equations similar to the Navier-Stokes equations.

Smagorinsky Model for the Sub-Grid Scale Eddy Viscosity

Among many others, Joseph Smagorinsky (1964) proposed a useful formula for the eddy viscosity in numerical models, based on the local derivatives of the velocity field and the local grid size:

$$\nu_t = \Delta x \Delta y \sqrt{\left(\frac{\partial u}{\partial x}\right)^2 + \left(\frac{\partial v}{\partial y}\right)^2 + \frac{1}{2}\left(\frac{\partial u}{\partial y} + \frac{\partial v}{\partial x}\right)^2}$$

Spalart–Allmaras, k–ε and k–ω Models

The Boussinesq hypothesis is employed in the Spalart–Allmaras (S–A), *k*–ε (*k*–epsilon), and *k*–ω (*k*–omega) models and offers a relatively low cost computation for the turbulence viscosity . The S–A model uses only one additional equation to model turbulence viscosity transport, while the k models use two.

Common Models

The following is a list of commonly employed models in modern engineering applications.

- Spalart–Allmaras (S–A)
- k–ε (k–epsilon)
- k–ω (k–omega)
- SST (Menter's Shear Stress Transport)

Spalart–Allmaras Turbulence Model

The Spalart–Allmaras model is a one equation model for turbulent viscosity. It solves a transport equation for a viscosity-like variable $\tilde{\nu}$. This may be referred to as the Spalart–Allmaras variable.

Original Model

The turbulent eddy viscosity is given by

$$\nu_t = \tilde{\nu} f_{v1}, \quad f_{v1} = \frac{\chi^3}{\chi^3 + C_{v1}^3}, \quad \chi := \frac{\tilde{\nu}}{\nu}$$

$$\frac{\partial \tilde{\nu}}{\partial t} + u_j \frac{\partial \tilde{\nu}}{\partial x_j} = C_{b1}[1 - f_{t2}]\tilde{S}\tilde{\nu} + \frac{1}{\sigma}\{\nabla \cdot [(\nu + \tilde{\nu})\nabla \tilde{\nu}] + C_{b2} |\nabla \nu|^2\} - \left[C_{w1} f_w - \frac{C_{b1}}{\kappa^2} f_{t2}\right]\left(\frac{\tilde{\nu}}{d}\right)^2 + f_{t1}\Delta U^2$$

$$\tilde{S} \equiv S + \frac{\tilde{\nu}}{\kappa^2 d^2} f_{v2}, \quad f_{v2} = 1 - \frac{\chi}{1 + \chi f_{v1}}$$

$$f_w = g\left[\frac{1 + C_{w3}^6}{g^6 + C_{w3}^6}\right]^{1/6}, \quad g = r + C_{w2}(r^6 - r), \quad r \equiv \frac{\tilde{\nu}}{\tilde{S}\kappa^2 d^2}$$

$$f_{t1} = C_{t1} g_t \exp\left(-C_{t2} \frac{\omega_t^2}{\Delta U^2}[d^2 + g_t^2 d_t^2]\right)$$

$$f_{t2} = C_{t3} \exp\left(-C_{t4}\chi^2\right)$$

$$S = \sqrt{2\Omega_{ij}\Omega_{ij}}$$

The rotation tensor is given by

$$\Omega_{ij} = \frac{1}{2}(\partial u_i / \partial x_j - \partial u_j / \partial x_i)$$

and d is the distance from the closest surface.

The constants are

$$\begin{aligned}
\sigma &= 2/3 \\
C_{b1} &= 0.1355 \\
C_{b2} &= 0.622 \\
\kappa &= 0.41 \\
C_{w1} &= C_{b1}/\kappa^2 + (1+C_{b2})/\sigma \\
C_{w2} &= 0.3 \\
C_{w3} &= 2 \\
C_{v1} &= 7.1 \\
C_{t1} &= 1 \\
C_{t2} &= 2 \\
C_{t3} &= 1.1 \\
C_{t4} &= 2
\end{aligned}$$

Modifications to Original Model

According to Spalart it is safer to use the following values for the last two constants:

$$\begin{aligned}
C_{t3} &= 1.2 \\
C_{t4} &= 0.5
\end{aligned}$$

Other models related to the S-A model:

DES (1999) [1]

DDES (2006)

Model for Compressible Flows

There are two approaches to adapting the model for compressible flows. In the first approach the turbulent dynamic viscosity is computed from

$$\mu_t = \rho\tilde{\nu}f_{v1}$$

where ρ is the local density. The convective terms in the equation for $\tilde{\nu}$ are modified to

$$\frac{\partial\tilde{\nu}}{\partial t} + \frac{\partial}{\partial x_j}(\tilde{\nu}u_j) = \text{RHS}$$

where the right hand side (RHS) is the same as in the original model.

Boundary Conditions

Walls: $\tilde{\nu} = 0$

Freestream

Ideally $\tilde{\nu} = 0$, but some solvers can have problems with a zero value, in which case $\tilde{\nu} <= \frac{\nu}{2}$ can be used.

This is if the trip term is used to "start up" the model. A convenient option is to set $\tilde{\nu} = 5\nu$ in the freestream. The model then provides "Fully Turbulent" behaviour, i.e., it becomes turbulent in any region that contains shear.

Chapter 8

Atmospheric Dispersion Modelling

Atmospheric dispersion modelling is the mathematical simulation of how air pollutants disperse in the ambient atmosphere. It is performed with computer programs that solve the mathematical equations and algorithms which simulate the pollutant dispersion. The dispersion models are used to estimate or to predict the downwind concentration of air pollutants or toxins emitted from sources such as industrial plants, vehicular traffic or accidental chemical releases.

Such models are important to governmental agencies tasked with protecting and managing the ambient air quality. The models are typically employed to determine whether existing or proposed new industrial facilities are or will be in compliance with the National Ambient Air Quality Standards (NAAQS) in the United States and other nations. The models also serve to assist in the design of effective control strategies to reduce emissions of harmful air pollutants. During the late 1960s, the Air Pollution Control Office of the U.S. EPA initiated research projects that would lead to the development of models for the use by urban and transportation planners. A major and significant application of a roadway dispersion model that resulted from such research was applied to the Spadina Expressway of Canada in 1971.

Air dispersion models are also used by public safety responders and emergency management personnel for emergency planning of accidental chemical releases. Models are used to determine the consequences of accidental releases of hazardous or toxic materials, Accidental releases may result fires, spills or explosions that involve hazardous materials, such as chemicals or radionuclides. The results of dispersion modelling, using worst case accidental release source terms and meteorological conditions, can provide an estimate of location impacted areas, ambient concentrations, and be used to determine protective actions appropriate in the event a release occurs. Appropriate protective actions may include evacuation or shelter in place for persons in the downwind direction. At industrial facilities, this type of consequence assessment or emergency planning is required under the Clean Air Act (United States) (CAA) codified in Part 68 of Title 40 of the Code of Federal Regulations.

The dispersion models vary depending on the mathematics used to develop the model, but all require the input of data that may include:

- Meteorological conditions such as wind speed and direction, the amount of atmospheric turbulence (as characterized by what is called the "stability class"), the ambient air temperature, the height to the bottom of any inversion aloft that may be present, cloud cover and solar radiation.
- Source term (the concentration or quantity of toxins in emission or accidental release source terms) and temperature of the material
- Emissions or release parameters such as source location and height, type of source (i.e., fire, pool or vent stack) and exit velocity, exit temperature and mass flow rate or release rate.
- Terrain elevations at the source location and at the receptor location(s), such as nearby homes, schools, businesses and hospitals.
- The location, height and width of any obstructions (such as buildings or other structures) in the path of the emitted gaseous plume, surface roughness or the use of a more generic parameter "rural" or "city" terrain.

Many of the modern, advanced dispersion modelling programs include a pre-processor module for the input of meteorological and other data, and many also include a post-processor module for graphing the output data and/or plotting the area impacted by the air pollutants on maps. The plots of areas impacted may also include isopleths showing areas of minimal to high concentrations that define areas of the highest health risk. The isopleths plots are useful in determining protective actions for the public and responders.

The atmospheric dispersion models are also known as atmospheric diffusion models, air dispersion models, air quality models, and air pollution dispersion models.

Atmospheric Layers

Discussion of the layers in the Earth's atmosphere is needed to understand where airborne pollutants disperse in the atmosphere. The layer closest to the Earth's surface is known as the *troposphere*. It extends from sea-level to a height of about 18 km and contains about 80 percent of the mass of the overall atmosphere. The *stratosphere* is the next layer and extends from 18 km to about 50 km. The third layer is the *mesosphere* which extends from 50 km to about 80 km. There are other layers above 80 km, but they are insignificant with respect to atmospheric dispersion modelling.

The lowest part of the troposphere is called the *atmospheric boundary layer (ABL)* or the *planetary boundary layer (PBL)* and extends from the Earth's surface to about 1.5 to 2.0 km in height. The air temperature of the atmospheric boundary layer decreases with increasing altitude until it reaches what is called the *inversion layer* (where the temperature increases with increasing altitude) that caps the atmospheric boundary layer. The upper part of the troposphere (i.e., above the inversion layer) is called the *free troposphere* and it extends up to the 18 km height of the troposphere.

The ABL is of the most important with respect to the emission, transport and dispersion of airborne pollutants. The part of the ABL between the Earth's surface and the bottom of the inversion layer is known as the mixing layer. Almost all of the airborne pollutants emitted into the ambient atmosphere are transported and dispersed within the mixing layer. Some of the emissions penetrate the inversion layer and enter the free troposphere above the ABL.

In summary, the layers of the Earth's atmosphere from the surface of the ground upwards are: the ABL made up of the mixing layer capped by the inversion layer; the free troposphere; the stratosphere; the mesosphere and others. Many atmospheric dispersion models are referred to as *boundary layer models* because they mainly model air pollutant dispersion within the ABL. To avoid confusion, models referred to as *mesoscale models* have dispersion modelling capabilities that extend horizontally up to a few hundred kilometres. It does not mean that they model dispersion in the mesosphere.

Gaussian Air Pollutant Dispersion Equation

The technical literature on air pollution dispersion is quite extensive and dates back to the 1930s and earlier. One of the early air pollutant plume dispersion equations was derived by Bosanquet and Pearson. Their equation did not assume Gaussian distribution nor did it include the effect of ground reflection of the pollutant plume. Sir Graham Sutton derived an air pollutant plume dispersion equation in 1947 which did include the assumption of Gaussian distribution for the vertical and crosswind dispersion of the plume and also included the effect of ground reflection of the plume.

Under the stimulus provided by the advent of stringent environmental control regulations, there was an immense growth in the use of air pollutant plume dispersion calculations between the late 1960s and today. A great many computer programs for calculating the dispersion of air pollutant emissions were developed during that period of time and they were called "air dispersion models". The basis for most of those models was the Complete Equation For Gaussian Dispersion Modelling Of Continuous, Buoyant Air Pollution Plumes shown below:

$$C = \frac{Q}{u} \cdot \frac{f}{\sigma_y \sqrt{2\pi}} \cdot \frac{g_1 + g_2 + g_3}{\sigma_z \sqrt{2\pi}}$$

where:

---

f = crosswind dispersion parameter

$$= \exp\left[-y^2/(2\sigma_y^2)\right]$$

g = vertical dispersion parameter = $g_1 + g_2 + g_3$

g_1 = vertical dispersion with no reflections

$$= \exp\left[-(z-H)^2/(2\sigma_z^2)\right]$$

g_2 = vertical dispersion for reflection from the ground

$$= \exp\left[-(z+H)^2/(2\sigma_z^2)\right]$$

g_3 = vertical dispersion for reflection from an inversion aloft

$$= \sum_{m=1}^{\infty} \{\exp\left[-(z-H-2mL)^2/(2\sigma_z^2)\right]$$

$$+\exp\left[-(z+H+2mL)^2/(2\sigma_z^2)\right]$$

$$+\exp\left[-(z+H-2mL)^2/(2\sigma_z^2)\right]$$

$$+\exp\left[-(z-H+2mL)^2/(2\sigma_z^2)\right]\}$$

C	= concentration of emissions, in g/m³, at any receptor located: x metres downwind from the emission source point y metres crosswind from the emission plume centreline z metres above ground level
Q	= source pollutant emission rate, in g/s
u	= horizontal wind velocity along the plume centreline, m/s
H	= height of emission plume centreline above ground level, in m
σ_z	= vertical standard deviation of the emission distribution, in m
σ_y	= horizontal standard deviation of the emission distribution, in m
L	= height from ground level to bottom of the inversion aloft, in m
exp	= the exponential function

The above equation not only includes upward reflection from the ground, it also includes downward reflection from the bottom of any inversion lid present in the atmosphere.

The sum of the four exponential terms in g_3 converges to a final value quite rapidly. For most cases, the summation of the series with $m = 1$, $m = 2$ and $m = 3$ will provide an adequate solution.

σ_z and σ_y are functions of the atmospheric stability class (i.e., a measure of the turbulence in the ambient atmosphere) and of the downwind distance to the receptor. The two most important variables affecting the degree of pollutant emission dispersion obtained are the height of the emission source point and the degree of atmospheric turbulence. The more turbulence, the better the degree of dispersion.

The resulting calculations for air pollutant concentrations are often expressed as an air pollutant concentration contour map in order to show the spatial variation in contaminant levels over a wide area under study. In this way the contour lines can overlay sensitive receptor locations and reveal the spatial relationship of air pollutants to areas of interest.

Whereas older models rely on stability classes for the determination of σ_y and σ_z, more recent models increasingly rely on the Monin-Obukhov similarity theory to derive these parameters.

Briggs Plume Rise Equations

The Gaussian air pollutant dispersion equation (discussed above) requires the input of H which is the pollutant plume's centreline height above ground level—and H is the sum of H_s (the actual physical height of the pollutant plume's emission source point) plus ΔH (the plume rise due the plume's buoyancy).

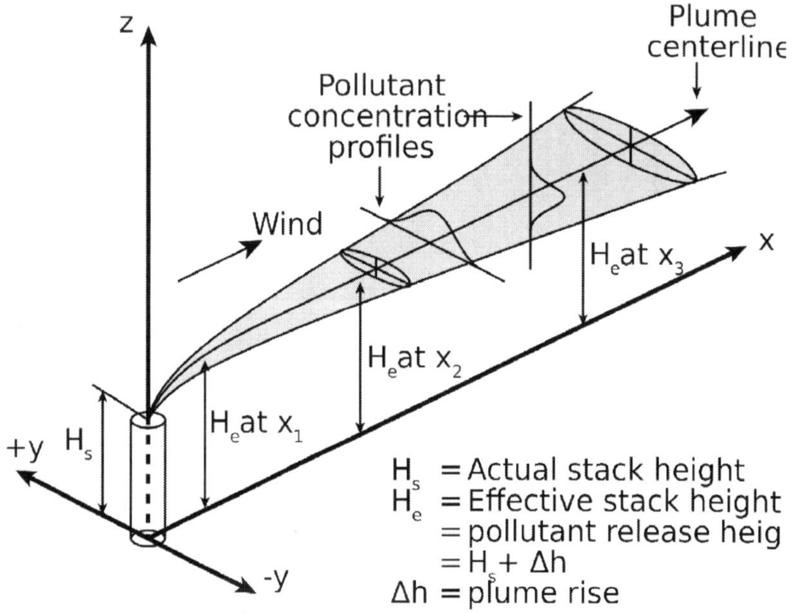

***Figure:** Visualization of a buoyant Gaussian air pollutant dispersion plume*

To determine ΔH, many if not most of the air dispersion models developed between the late 1960s and the early 2000s used what are known as "the Briggs equations." G.A. Briggs first published his plume rise observations and comparisons in 1965. In 1968, at a symposium sponsored by CONCAWE (a Dutch organisation), he compared many of the plume rise models then available in the literature. In that same year, Briggs also wrote the section of the publication edited by Slade dealing with the comparative analyses of plume rise models. That was followed in 1969 by his classical critical review of the entire plume rise literature, in which he proposed a set of plume rise equations which have become widely known as "the Briggs equations". Subsequently, Briggs modified his 1969 plume rise equations in 1971 and in 1972.

Briggs divided air pollution plumes into these four general categories:

- Cold jet plumes in calm ambient air conditions
- Cold jet plumes in windy ambient air conditions

- Hot, buoyant plumes in calm ambient air conditions
- Hot, buoyant plumes in windy ambient air conditions

Briggs considered the trajectory of cold jet plumes to be dominated by their initial velocity momentum, and the trajectory of hot, buoyant plumes to be dominated by their buoyant momentum to the extent that their initial velocity momentum was relatively unimportant. Although Briggs proposed plume rise equations for each of the above plume categories, *it is important to emphasize that "the Briggs equations" which become widely used are those that he proposed for bent-over, hot buoyant plumes.*

In general, Briggs's equations for bent-over, hot buoyant plumes are based on observations and data involving plumes from typical combustion sources such as the flue gas stacks from steam-generating boilers burning fossil fuels in large power plants.

Therefore the stack exit velocities were probably in the range of 20 to 100 ft/s (6 to 30 m/s) with exit temperatures ranging from 250 to 500 °F (120 to 260 °C).

A logic diagram for using the Briggs equations to obtain the plume rise trajectory of bent-over buoyant plumes is presented below:

LOGIC DIAGRAM FOR BRIGGS' EQUATIONS TO CALCULATE THE RISE OF A BUOYANT PLUME

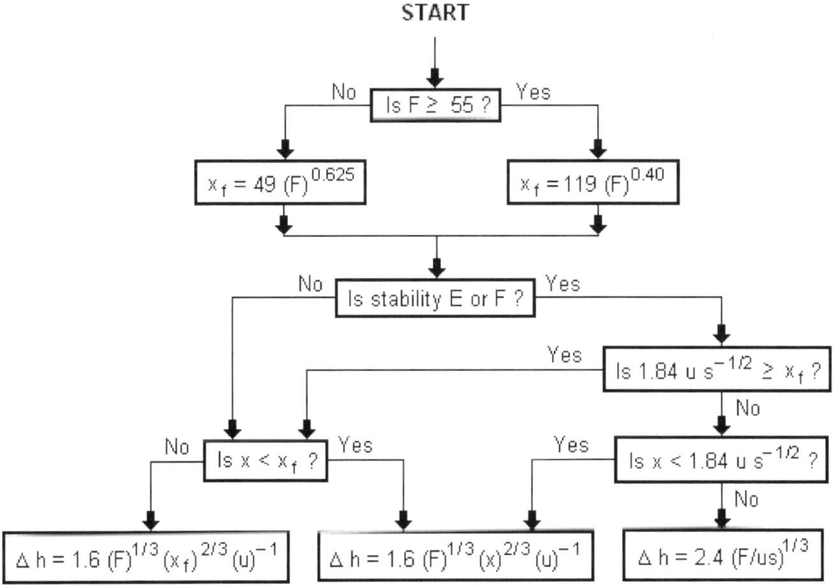

where:

Δh = plume rise, in m

F = buoyancy factor, in m^4s^{-3}

x = downwind distance from plume source, in m

x_f = downwind distance from plume source to point of maximum plume rise, in m

u = windspeed at actual stack height, in m/s

s = stability parameter, in s^{-2}

The above parameters used in the Briggs' equations are discussed in Beychok's book.

Breakwater (Structure)

Breakwaters are structures constructed on coasts as part of coastal defence or to protect an anchorage from the effects of both weather and longshore drift.

Purposes of Breakwaters

Offshore breakwaters, also called bulkhead, reduce the intensity of wave action in inshore waters and thereby reduce coastal erosion or provide safe harborage. Breakwaters may also be small structures designed to protect a gently sloping beach and placed one to three hundred feet offshore in relatively shallow water.

An anchorage is only safe if ships anchored there are protected from the force of high winds and powerful waves by some large underwater barrier which they can shelter behind. Natural harbours are formed by such barriers as headlands or reefs. Artificial harbors can be created with the help of breakwaters. Mobile harbours, such as the D-Day Mulberry harbours, were floated into position and acted as breakwaters. Some natural harbours, such as those in Plymouth Sound, Portland Harbour and Cherbourg, have been enhanced or extended by breakwaters made of rock.

Unintended Consequences

The dissipation of energy and relative calm water created in the lee of the breakwaters often encourage accretion of sediment (as per the design of the breakwater scheme). However this can lead to excessive salient build up, leading to tombolo formation reducing longshore drift shoreward of the breakwaters (Sea Palling, UK). This trapping of sediment can cause adverse effects down drift of the breakwaters leading to beach sediment starvation and increased erosion. This may then lead to further engineering protection being needed down drift of the breakwater development.

Breakwaters are subject to damage, and overtopping in severe storms events.

Construction

Breakwaters can be constructed with one end linked to the shore, in which case they are usually classified as sea walls; otherwise they are positioned offshore from as little as 100m up to 300-600m from the original shoreline. There are two main types of offshore breakwater, single and multiple; single as the name suggests means the breakwater consists of one unbroken barrier, which multiple breakwaters (in numbers anywhere from 2-20) are positioned with gaps in between (50-300m). Length of gap is largely governed by the interacting wavelengths. Breakwaters may be either fixed or floating, and impermeable or permeable to allow sediment transfer shoreward of the structures, the choice depending tidal range and water depth. They usually consist of large pieces of rock (granite) weighting up to 16 tonnes each or rubble-mound. Their design is influenced by the angle of wave approach and other environmental parameters. Breakwater construction can be either parallel or perpendicular to the coast, depending on the shoreline requirements.

Types of Breakwater Structures

A breakwater structure is designed to absorb the energy of the waves that hit it, either by using mass (e.g., with caissons), or by using a revetment slope (e.g., with rock or concrete armour units).

In Coastal Engineering, a revetment is a land backed structure whilst a breakwater is a sea backed structure (i.e., water on both sides).

Caisson breakwaters typically have vertical sides and are usually erected where it is desirable to berth one or more vessels on the inner face of the breakwater. They use the mass of the caisson and the fill within it to resist the overturning forces applied by waves hitting them. They are relatively expensive to construct in shallow water, but in deeper sites they can offer a significant saving over revetment breakwaters.

Rubble mound breakwaters use structural voids to dissipate the wave energy. Rock or concrete armour units on the outside of the structure absorb most of the energy, while gravels or sands prevent the wave energy's continuing through the breakwater core. The slopes of the revetment are typically between 1:1 and 1:2, depending upon the materials used. In shallow water, revetment breakwaters are usually relatively inexpensive. As water depth increases, the material requirements, and hence costs, increase significantly.

Advanced Numerical Study

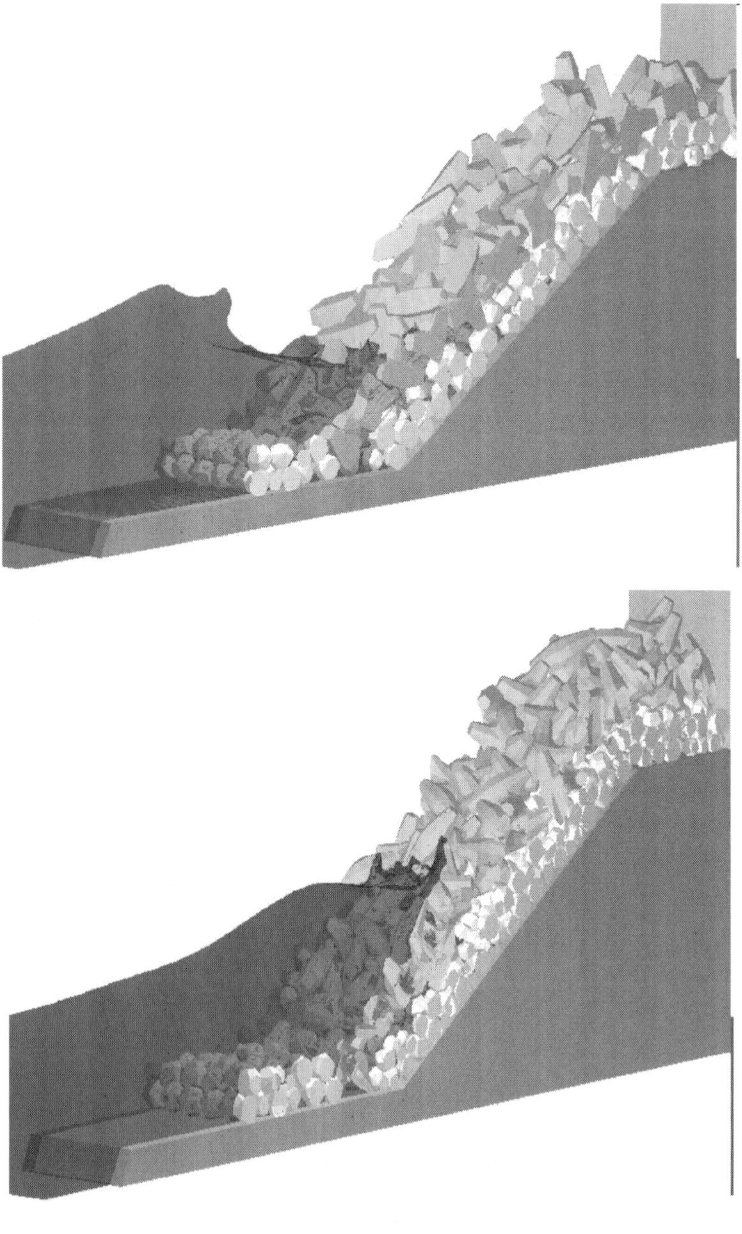

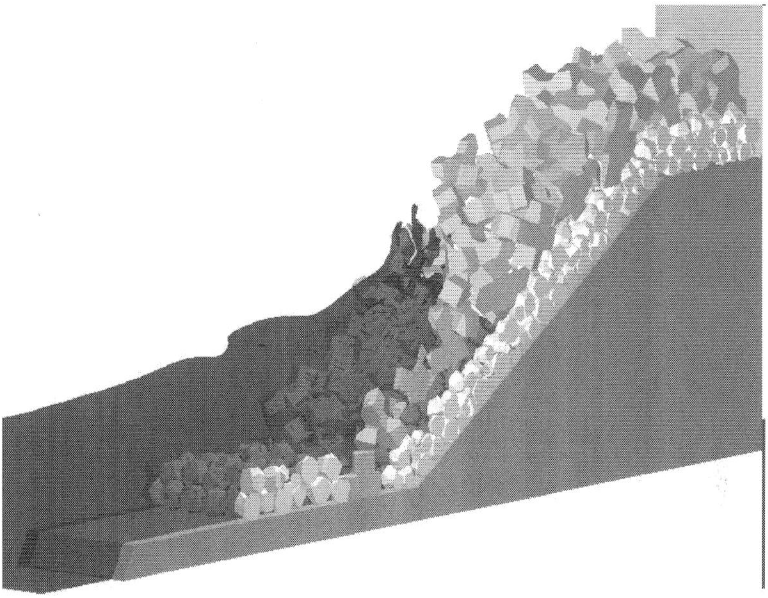

Figure: *3D Numerical Simulation - MEDUS 2009*

The Maritime Engineering Division of the University of Salerno (MEDUS) developed a new procedure for studying in greater detail the interactions between maritime breakwaters (submerged or emerged) and the waves that hit them by making integrated use of CAD and CFD software.

In the numerical simulations, the filtration motion of the fluid within the interstices, which normally exist in a breakwater, is estimated by integrating the RANS equations, coupled with a RNG turbulence model inside the voids, instead of using classical equations for porous media.

The breakwaters were modelled, in analogy to full size construction or physical laboratory tests, by overlapping three-dimensional elements and having the numerical grid thickened in order to have some computational nodes along the flow paths among the breakwater's blocks.

Beach Nourishment

Beach nourishment— also referred to as beach replenishment or sand replenishment —describes a process by which sediment (usually sand) lost through longshore drift or erosion is replaced from sources outside of the eroding beach. A wider beach can reduce storm damage to coastal structures by dissipating energy across the surf zone, protecting upland structures and infrastructure from storm surges, tsunamis and unusually high tides. Beach nourishment is typically part of a larger coastal defence scheme. Nourishment is typically a repetitive process, since it does not remove the physical forces that cause erosion, but simply mitigates their effects.

The first nourishment project in the U.S. was at Coney Island, New York in 1922-23 and is now a common shore protection measure utilised by public and private entities.

Nourishment is one of three commonly accepted methods for protecting shorelines. The structural alternative involves constructing a seawall, revetment, groin or breakwater. Alternatively, with "managed retreat" the shoreline is left to erode, while relocating buildings and infrastructure further inland. Nourishment gained popularity because it preserved beach resources and avoided the negative effects of hard structures. Instead, nourishment creates a "soft" (i.e., non-permanent) structure by creating a larger sand reservoir, pushing the shoreline seaward.

Causes of Erosion

Beaches can erode both naturally and due to impact of humans. Erosion is a natural response to storm activity. During storms, sand from the visible beach submerges to form storm bars that protect the beach. Submersion is only part of the cycle. During calm weather smaller waves return sand from the storm bar to the visible beach surface in a process called accretion. The term *erosion* conjures visions of environmental damage so the term submersion often replaces it in describing a healthy sandy beach. Continental drift erodes coastlines naturally. Ocean currents can change, disrupting the submersion/ accretion cycle.

Some beaches do not have enough sand available to coastal processes to respond naturally to storms. When there is not enough sand left available on a beach, then there is no recovery of the beach following storms.

Many areas of high erosion are due to human activities. Reasons can include seawalls locking up sand dunes, coastal structures like ports and harbors that prevent longshore drift, dams and other river management structures. These activities interfere with the natural sediment flows either through dam construction (thereby reducing riverine sediment sources) or construction of littoral barriers such as jetties, or by deepening of inlets; thus preventing longshore transport of sediment across these channels.

Visible and Submerged Sand

The distinction between total sand in a beach and the proportion of the sand above the waterline (submersion fraction) critically impacts beach nourishment. Two beaches with the same amount of visible sand may look much different under water. An eroded beach with substantial submerged sand surrounding it may recover without nourishment. Nourishing a beach that has little submerged sand requires addressing the reason that the submerged sand is missing. Otherwise, the same forces that stripped the submerged sand once are likely to do so again.

The amount of submerged sand eroded is typically much greater than the amount of missing sand on shore. Replacing only the visible sand is insufficient without replacing the sand that once supported the accretion part of the process. In that circumstance, the beach is unstable and the visible sand quickly erodes. If human activity is a major cause of the erosion, mitigating that activity may be more cost effective over both short and long term periods than nourishment.

Requirements for Effective Nourishment

Sediment texture (grain size and sorting) is critical for success. Sand fill must be compatible with native beach sand. In some cases, beaches have been nourished using a finer sand than the original. Thermoluminescence monitoring reveals that storms can erode such beaches far more quickly than the natural beach. This was observed at the Waikiki nourishment project in Hawaii.

Profile Nourishment

Beach Profile Nourishment describes programs that nourish the full beach profile. In this instance, "profile" means the slope of the uneroded beach from above the water to well out to sea, not just the visible portion. The Gold Coast profile nourishment program placed 75% of its total sand volume below low water level. Some coastal authorities *overnourish* the below water beach (aka "nearshore nourishment") so that over time the natural beach increases in size. These approaches do not permanently protect beaches eroded by human activity. Doing so still requires mitigating that activity.

Replenishment Material

The selection of suitable material for a particular project depends upon the design needs, environmental factors, transport costs considering both short and long-term implications.

The most important material characteristic is the sediment's grain size, which must closely match the native material. Excess silt and clay fraction (mud) versus the natural turbidity in the nourishment area disqualifies some materials.

Projects that did not match grain sizes performed relatively poorly. Nourishment sand that is only slightly smaller than native sand can result in significantly narrower equilibrated dry beach widths compared to sand the same size as (or larger than) native sand. Evaluating material fit requires a sand survey that usually includes geophysical profiles and surface and core samples.

Type	*Description*	*Environmental issues*
Offshore	Exposure to open sea makes this the most difficult operational environment. Must consider the effects of altering depth on wave energy at the shoreline. May be combined with a navigation project.	Impacts on hard bottom and migratory species.
Inlet	Sand between jetties in a stabilized inlet. Often associated with dredging of navigational channels and the ebb- or flood-tide deltas of both natural and jettied inlets.	
Accretional Beach	Generally not suitable because of damage to source beach.	
Upland	Generally the easiest to obtain permits and assess impacts from a land source. Offers opportunities for mitigation. Limited quantity and quality of economical deposits.	Potential secondary impacts from mining and overland transport.
Riverine	Potentially high quality and sizeable quantity. Transport distance a possible cost factor.	May interrupt natural coastal sand supply.
Lagoon	Often excessively fine grained. Often close to barrier beaches and in sheltered waters, easing construction. Principal sources are flood-tide deltas.	Can compromise wetlands.
Artificial or non-indigenous	Typically, high transport and redistribution costs. Some laboratory experiments done on recycling broken glass. Aragonite from Bahamas a possible source.	
Emergency	Deposits near inlets and local sinks and sand from stable beaches with adequate supply. Generally used only following a storm or given no other affordable option. May be combined with a navigation project.	Harm to source site. Poor match to target requirements.

'Advantages

1. widens the recreational beach.
2. Structures behind beach are protected as long as the added sand remains.

Disadvantages:

1. Beach nourishment sand often (in fact, usually) erodes faster than the natural sand on the beach.

2. Beach nourishment is expensive, and must be repeated periodically.
3. The beach turns into a construction zone during nourishment.
4. The process of nourishment may damage, destroy or otherwise hurt marine and beach life by burying it.
5. The sand added to the beach is often different from the natural beach sand. It can be hard to find a perfect match.

Environmental Issues

Beach nourishment has significant impacts on local ecosystems. Nourishment may cause direct mortality to sessile organisms in the target area by burying them in the new sand. Seafloor habitat in both source and target areas are disrupted, e.g., when sand is deposited on coral reefs or when deposited sand hardens. Imported sand may differ in character (chemical makeup, grain size, non-native species) from that of the target environment. Reduced light availability, affecting nearby reefs and submerged aquatic vegetation. Imported sand may contain material toxic to local species. Removing material from near-shore environments may destabilize the shoreline, in part by steepening its submerged slope. Related attempts to reduce future erosion may provide a false sense of security that increases development pressure.

Sea Turtles

Newly-deposited sand can harden and complicate nest-digging for turtles. However, nourishment can provide more/better habitat for them, as well as for sea birds and beach flora. Florida addressed the concern that dredge pipes would suck turtles into the pumps by adding a special grill to the dredge pipes that ensured that turtles were not sucked in with the sand.

Alternatives/Complements to Nourishment

Nourishment is not the only technique used to address eroding beaches. Others can be used singly or in combination with nourishment, driven by economic, environmental, and political considerations. Human activities such as dam construction can interfere with natural sediment flows (thereby reducing riverine sediment sources.) Construction of littoral barriers such as jetties and deepening of inlets can prevent longshore sediment transport.

Structural

The structural approach attempts to prevent erosion. Armoring involves building revetments, seawalls, detached breakwaters, groins, etc. Structures that run parallel to the shore (seawalls or revetments) prevent erosion. While this protects structures, it doesn't not protect the beach which is outside the wall. The beach disappears over a period that ranges from months to decades.

Groynes and breakwaters that run perpendicular to the shore protect the shore from erosion. Filling a breakwater with imported sand can stop the breakwater from trapping sand from the littoral stream (the ocean running along the shore.) Otherwise the breakwater may deprive downstream beaches of sand and accelerate erosion there. Armoring may restrict beach/ocean access, enhance erosion of adjacent shorelines, and requires long-term maintenance.

Managed Retreat

A second option is to move structures and infrastructure inland as the shoreline erodes. Retreat is more often chosen in areas of rapid erosion and in the presence of little or obsolete development.

Recruitment

Appropriately constructed and sited fences can capture blowing sand, building/restoring sand dunes, and progressively protecting the beach from the wind, and the shore from blowing sand.

Beach Drainage

All beaches lose and gain sand depending on the tides, precipitation, wind, waves etc. Wet beaches tend to lose sand. Waves infiltrate dry beaches easily and deposit sandy sediment. Generally a beach is wet during falling tide, because the sea sinks faster than the beach drains. As a result most erosion happens during falling tide. Beach drainage (beach dewatering) using Pressure Equalizing Modules (PEMs) allow the beach to drain more effectively during falling tide. Fewer hours of wet beach translate to less erosion. Permeable PEM tubes inserted vertically into the foreshore connect the different layers of groundwater. The groundwater enters the PEM tube allowing gravity to conduct it to a coarser sand layer, where it can drain more quickly. The PEM modules are placed in a row from the dune to the mean low waterline. Distance between rows is typically 300 feet (91 m) but this is project-specific. PEM systems come in different sizes. Modules connect layers with varying hydraulic conductivity. Air/water can enter and equalize pressure.

PEMs are minimially invasive, typically covering approximately 0.00005% of the beach. The effects are local, and bring no known harm to flora or fauna, including nesting sea turtles. The tubes are below the beach surface, with no visible impact. Installation takes little time. Costs are low, because PEMs require few materials, and no energy-intensive dredging. PEM installations have been installed on beaches in Denmark, Sweden, Malaysia, and Florida, USA. The effectiveness of beach dewatering, however, is debatable and has not been proven convincingly on life-size beaches.

Costs

Nourishment is typically a repetitive process, since nourishment does not remove the physical forces that cause erosion; it simply mitigates their effects. A benign environment increases the interval between nourishment projects, reducing costs. Conversely, high erosion rates may render nourishment financially impractical.

In many coastal areas, the economic impacts of a wide beach can be substantial. The 10 miles (16 km)–long shoreline fronting Miami Beach, Florida was replenished over the period 1976–1981. The project cost approximately $64,000,000 and revitalized the area's economy. Prior to nourishment, in many places the beach was too narrow to walk along, especially during high tide.

History

The first nourishment project in the U.S. was constructed at Coney Island, New York in 1922-1923.

Nourishment Projects

The setting of a beach nourishment project is key to design and potential performance. Possible settings include a long straight beach, an inlet that may be either natural or modified, and a pocket beach. Rocky or seawalled shorelines, that otherwise have no sediment, present unique problems.

Cancun, Mexico

Federal and state governments in Mexico have invested about $71 million ($957 million pesos) throughout the state of Quintana Roo in restoring the beaches along Cancun, Playa del Carmen, and Cozumel.

Hurricane Wilma hit the beaches of Cancun and the Riviera Maya in 2005. The initial nourishment project was unsuccessful, leading to a second round that began in September 2009, and was scheduled to complete in early 2010. The project designers and the government committed to invest in beach maintenance to address future erosion. Project designers considered factors such as the time of year and sand characteristics such as density. Restoration in Cancun was expected to deliver 1.3 billion US gallons (4,900,000 m^3) of sand to replenish 450 metres (1,480 ft) of coastline. This time, the beach is promised to last at least 10 years.

Northern Gold Coast, Queensland, Australia

Gold Coast beaches in Queensland, Australia have experienced periods of severe erosion. In 1967 a series of 11 cyclones removed most of the sand from Gold Coast beaches. The Government of Queensland engaged engineers from Delft University in the Netherlands to advise them. The 1971 Delft Report outlined a series of works for Gold Coast Beaches, including beach nourishment and an artificial reef. By 2005 most of the recommendations had been implemented.

The Northern Gold Coast Beach Protection Strategy (NGCBPS) was a A$10 million investment. NGCBPS was developed between 1992 and 1999 and the works were completed between 1999 and 2003. The project included dredging 3,500,000 cubic metres (4,600,000 cu yd) of compatible sand from the Gold Coast Broadwater and delivering it through a pipeline to nourish 5 kilometres (3.1 mi) between of beach between Surfers Paradise and Main Beach.

The new sand was stabilized by an artificial reef constructed at Narrowneck out of huge geotextile sand bags. The new reef was designed to improve wave conditions for surfing. A key monitoring program for the NGCBPS is the ARGUS coastal camera system operated by the University of New South Wales.

The cost/benefit ratio for NGCBPS was conservatively estimated at 75:1 for a A$10 million investment into beach replenishment. The benefits were estimated from a model of lost visitor nights in hotels following previous erosion events. NGCBPS so improved beach health that recovery following minor and moderate storms occurred within weeks. Additional unquantified benefits included lifestyle benefits for residents, additional public open space and improved fishing, diving and surfing conditions.

Netherlands

More than one-quarter of the Netherlands is below sea level and about 81% of the coast consists of sand dune or beach. The shoreline is closely monitored by yearly recording of the cross section at points 250 metres (820 ft) apart, to ensure adequate protection. Where long-term erosion is identified, beach nourishment using high-capacity suction dredgers is deployed.

Hawaii

Waikiki: Hawaii planned to replenish Waikiki beach in 2010. Budgeted at $2.5 million, the project covered 1,700 feet (520 m) in an attempt to return the beach to its 1985 width. Prior opponents supported this project, because the sand was to come from nearby shoals, reopening a blocked channel and leaving the overall local sand volume unchanged, while closely matching the "new" sand to existing materials. The project planned to apply up to 24,000 cubic yards (18,000 m^3) of sand from deposits located 1,500 to 3,000 feet (460 to 910 m) offshore at a depth of 10 to 20 feet (3.0 to 6.1 m). The project was larger than the prior recycling effort in 2006-07, which moved 10,000 cubic yards (7,600 m^3).

Maui: Maui, Hawaii illustrated the complexities of even small-scale nourishment projects. A project at Sugar Cove transported upland sand to the beach. The sand allegedly was finer than the original sand and contained excess silt that enveloped coral heads, smothering the coral and killing small animals that lived in and around it. As in other projects, on-shore sand availability was limited, forcing consideration of more expensive offshore sources.

A second project, along Stable Road, that attempted to slow rather than halt erosion, was stopped halfway toward its goal of adding 10,000 cubic yards (7,600 m^3) of sand. The beaches had been retreating at a "comparatively fast rate" for half a century. The restoration was complicated by the presence of old seawalls, groins, piles of rocks and other structures.

This project used sand-filled Geotube groins placed that were originally to remain in place for up to 3 years. The pipeline was anchored by concrete blocks attahced by fibre straps. A video showed the blocks bouncing off the coral in the current, killing whatever they touched. In places the straps broke, allowing the pipe to move across the reef, "planing it down". Bad weather exacerbated the damaging movement and led to project shutdown. The smooth, cylindrical Geotubes could be difficult to climb over before they were covered by sand.

Supporters claimed that 2010's seasonal summer erosion was less than in prior years, although the beach was narrower after the restoration ended than in 2008. Authorities were studying whether to require the project to remove the groins immediately. Potential alternatives to Geotubes for moving sand included floating dredges and/or trucking in sand dredged offshore.

A final consideration was sea level rise and that Maui was sinking under its own weight. Both islands surround massive mountains (Haleakala, Mauna Loa, and Mauna Kea) and were expanding a giant dimple in the ocean floor, some 30,000 feet (9,100 m) below the mountain summits.

Hong Kong

The beach in Gold Coast was built as an artificial beach in the 1990s with HK$60m. Sands have to be supplied periodically, especially after typhoons, to keep the beach viable.

Measuring Project Impact

Nourishment projects usually involve physical, environmental and economic objectives. Typical physical measures include dry beach width, remaining post-storm sand volume, post-storm damage avoidance assessments and aqueous sand volume. Environmental measures include marine life distribution, habitat and population counts. Economic impacts include recreation, tourism, flood and "disaster" prevention. Techniques for incorporating nourishment projects into flood insurance costs and disaster assistance remain controversial. The ability to predict the performance of a beach nourishment project is best for a project constructed on a long, straight shoreline without the complications of inlets or engineered structures. In addition, predictability is better for overall performance, e.g., average shoreline change, rather than shoreline change at a specific location.

Nourishment can affect eligibility in the National Flood Insurance Program and federal disaster assistance.

Clapotis

In hydrodynamics, the clapotis (from French: *"lapping of water"*) is a non-breaking standing wave pattern, caused for example, by the reflection of a travelling surface wave train from a near vertical shoreline like a breakwater, seawall or steep cliff. The resulting clapotic wave does not travel horizontally, but has a fixed pattern of nodes and antinodes. These waves promote erosion at the toe of the wall, and can cause severe damage to shore structures. The term was coined in 1877 by French mathematician and physicist Joseph Valentin Boussinesq who called these waves 'le clapotis' meaning 'standing waves'.

In the idealized case of "full clapotis" where a purely monotonic incoming wave is completely reflected normal to a solid vertical wall, the standing wave height is twice the height of the incoming waves at a distance of one half wavelength from the wall. In this case, the circular orbits of the water particles in the deep-water wave are converted to purely linear motion, with vertical velocities at the antinodes, and horizontal velocities at the nodes. The standing waves alternately rise and fall in a mirror image pattern, as kinetic energy is converted to potential energy, and vice versa. In his 1907 text, *Naval Architecture*, Cecil Peabody described this phenomenon:

At any instant the profile of the water surface is like that of a trochoidal wave, but the profile instead of appearing to run to the right or left, will grow from a horizontal surface, attain a maximum development, and then flatten out till the surface is again horizontal; immediately another wave profile will form with its crests where the hollows formerly were, will grow and flatten out, etc. If attention is concentrated on a certain crest, it will be seen to grow to its greatest height, die away, and be succeeded in the same place by a hollow, and the interval of time between the successive formations of crests at a given place will be the same as the time of one of the component waves."

Related Phenomena

True clapotis is very rare, because the depth of the water or the precipitousness of the shore are unlikely to completely satisfy the idealized requirements. In the more realistic case of partial clapotis, where some of the incoming wave energy is dissipated at the shore, the incident wave is less than 100% reflected, and only a partial standing wave is formed where the water particle motions are elliptical. This may also occur at sea between two different wave trains of near equal wavelength moving in opposite directions, but with unequal amplitudes. In partial clapotis the wave envelope contains some vertical motion at the nodes. When a wave train strikes a wall at an oblique angle, the reflected wave train departs at the supplementary angle causing a cross-hatched wave interference pattern known as the clapotis gaufré ("waffled clapotis"). In this situation, the individual crests formed at the intersection of the incident and reflected wave train crests move parallel to the structure. This wave motion, when combined with the resultant vortices, can erode material from the seabed and transport it along the wall, undermining the structure until it fails.

Clapotic waves on the sea surface also radiate infrasonic microbaroms into the atmosphere, and seismic signals called microseisms coupled through the ocean floor to the solid Earth.

Iribarren Number

In fluid dynamics, the Iribarren number or Iribarren parameter – also known as the surf similarity parameter and breaker parameter – is a dimensionless parameter used to model several effects of (breaking) surface gravity waves on beaches and coastal structures. The parameter is named after the Spanish engineer Ramón Iribarren Cavanillas (1900–1967), who introduced it to describe the occurrence of wave breaking on sloping beaches.

For instance, the Iribarren number is used to describe breaking wave types on beaches; or wave run-up on – and reflection by – beaches, breakwaters and dikes.

Definition

The Iribarren number – often denoted as *Ir* or ξ – is defined as:

$$\xi = \frac{\tan\alpha}{\sqrt{H/L_0}}, \quad \text{with} \quad L_0 = \frac{g}{2\pi}T^2,$$

where ξ is the Iribarren number, α is the bed slope, H is the wave height, L_0 is the deep-water wavelength, T is the period and g is the gravitational acceleration. Depending on the application, different definitions of H and T are used, for example: for periodic waves the wave height H_0 at deep water or the breaking wave height H_b at the edge of the surf zone. Or, for random waves, the significant wave height H_s at a certain location.

Breaker Types

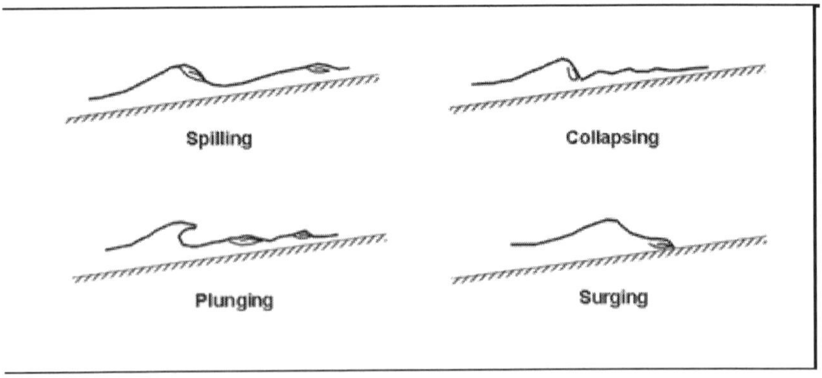

***Figure:** Breaker types.*

The type of breaking wave – spilling, plunging, collapsing or surging – depends on the Iribarren number. According to Battjes (1974), for periodic waves propagating on a plane beach, two possible choices for the Iribarren number are:

$$\xi_0 = \frac{\tan\alpha}{\sqrt{H_0 / L_0}} \quad \text{or} \quad \xi_b = \frac{\tan\alpha}{\sqrt{H_b / L_0}},$$

where H_0 is the offshore wave height in deep water, and H_b is the value of the wave height at the break point (where the waves start to break). Then the breaker types dependence on the Iribarren number (either ξ_0 or ξ_b) is approximately:

Breaker Type	ξ_0***-range***	ξ_b***-range***
surging or collapsing	$\xi_0 > 3.3$	$\xi_b > 2.0$
plunging	$0.5 < \xi_0 < 3.3$	$0.4 < \xi_b < 2.0$
spilling	$\xi_0 < 0.5$	$\xi_b < 0.4$

Hudson's Equation

Hudson's Equation, also known as Hudson's Formula, is an equation used by coastal engineers to calculate the minimum size of riprap (rock armour blocks) required to provide *satisfactory* stability characteristics for rubble structures such as breakwaters under attack from storm wave conditions.

The equation was developed by the United States Army Corps of Engineers, Waterways Experiment Station (WES), following extensive investigations by Hudson (1953, 1959, 1961a, 1961b).

Initial Equation

The equation itself is:

$$W = \frac{\gamma_r H^3}{K_D \Delta^3 \cot\theta}$$

where:

- W is the design weight of the riprap armour (Newton)
- γ_r is the specific weight of the armour blocks (N/m^3)
- H is the design wave height at the toe of the structure (m)
- K_D is a dimensionless stability coefficient, deduced from laboratory experiments for different kinds of armour blocks and for very small damage (a few blocks removed from the armour layer) (-):
- K_D = around 3 for natural quarry rock
- K_D = around 10 for artificial interlocking concrete blocks
- Δ is the dimensionless relative buoyant density of rock, i.e. $(\rho_r / \rho_w - 1)$ = around 1.58 for granite in sea water

- ρ_r and ρ_w are the densities of rock and (sea)water (-)
- θ is the angle of revetment with the horizontal

Updated Equation

This equation was rewritten as follows in the nineties:

$$\frac{H_s}{\Delta D_{n50}} = (K_D cot\theta)^{1/3}$$

where

- H_s is the design significant wave height at the toe of the structure (m)
- Δ is the dimensionless relative buoyant density of rock, i.e. $(\rho_r / \rho_w - 1)$ = around 1.58 for granite in sea water
- ρ_r and ρ_w are the densities of rock and (sea)water (-)
- D_{n50} is the nominal median diameter of armour blocks = $(W_{50}/\rho_r)^{1/3}$ (m)
- K_D is a dimensionless stability coefficient, deduced from laboratory experiments for different kinds of armour blocks and for very small damage (a few blocks removed from the armour layer) (-):
- K_D = around 3 for natural quarry rock
- K_D = around 10 for artificial interlocking concrete blocks
- θ is the angle of revetment with the horizontal

The armour blocks may be considered stable if the *stability number* $N_s = H_s / \Delta D_{n50} < 1.5$ to 2, with damage rapidly increasing for $N_s > 3$.

Obviously, these equations may be used for preliminary design, but scale model testing (2D in wave flume, and 3D in wave basin) is absolutely needed before construction is undertaken.

Revetment

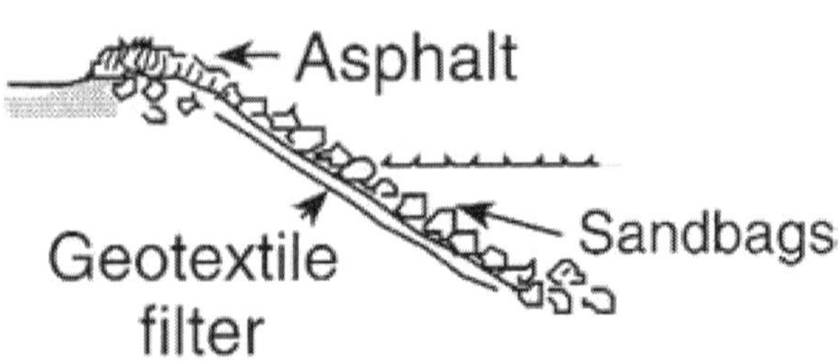

Figure: *Asphalt and sandbag revetment with a geotextile filter*

In stream restoration, river engineering or coastal management, revetments are sloping structures placed on banks or cliffs in such a way as to absorb the energy of incoming water. In military engineering they are structures, again sloped, formed to secure an area from artillery, bombing, or stored explosives. River or coastal revetments are usually built to preserve the existing uses of the shoreline and to protect the slope, as defence against erosion.

Freshwater Revetments

Any revetments are used to line the banks of freshwater rivers, lakes, and man-made reservoirs, especially to prevent damage during periods of floods or heavy seasonal rains. Many materials may be used: wooden piles, loose-piled boulders or concrete shapes, or more solid banks.

Revetments as Coastal Defence

Revetments are used as a low-cost solution for coastal erosion defence in areas where crashing waves may otherwise deplete the coastline. Wooden revetments are made of planks laid against wooden frames so that they disrupt the force of the water. Although once popular, the use of wooden revetments has largely been replaced by modern concrete-based defence structures such as tetrapods.

Tetrapods

In coastal engineering, a tetrapod is a four-legged concrete structure used as armour unit on breakwaters. The tetrapod's shape is designed to dissipate the force of incoming waves by allowing water to flow around rather than against it, and to reduce displacement by allowing a random distribution of tetrapods to mutually interlock. Tetrapods have been shown to offer little advantage compared to other concrete armour units. Layering and the amount of tetrapods have shown to have no appreciable effect on the stability of the tetrapod structures although stability increases with the size of the tetrapod pad. Tetrapods have also been criticized for causing more damage than they prevent because they alter ocean currents and disrupt the natural cycles of erosion and deposition that form and reshape coasts. Concrete coastal installations can also be lethally dangerous to swimmers and surfers, as well to shipping and recreational boaters.

Fortifications

According to the U.S. National Park Service, and referring mostly to their employment in the American Civil War, a revetment is defined as a "retaining wall constructed to support the interior slope of a parapet. Made of logs, wood planks, fence rails, fascines, gabions, hurdles, sods, or stones, the revetment provided additional protection from enemy fire, and, most importantly, kept the interior slope nearly vertical. Stone revetments commonly survive. A few log revetments have been preserved due to high resin pine or cypress and porous sandy soils. After an entrenchment was abandoned, many log or rail revetments were scavenged for other uses, causing the interior slope to slump more quickly. An interior slope will appear more vertical if the parapet eroded with the revetment still in place."

Sand Dune Stabilization

Sand dunes are common features of shoreline and desert environments. Dunes provide habitat for highly specialised plants and animals, including rare and endangered species. They can protect beaches from erosion and recruit sand to eroded beaches. Dunes are threatened by human activity, both intentional and unintentional. Countries such as the United States, Australia, Canada, New Zealand, the United Kingdom, and Netherlands, operate significant dune protection programs. Stabilizing dunes involves multiple actions. Planting vegetation reduces the impact of wind and water. Fences catch sand and other material. Footpaths protect dunes from damage from foot traffic.

Vegetation

The location of the dune limits the types of plant that can thrive there. Beach dunes consist of the *foredune*, the angled side which faces the ocean, the *sand plain* at the top of the dune, which may or may not be present, and the *backdune*, the angled side that faces away from the ocean.

Foredune Flora

Plants that thrive on the foredune must be tolerant to salt spray, strong winds, and burial by blowing/accumulating sand. Typical vegetation includes Ammophila arenaria, Honckenya peploides, Cakile maritima, and Spartina coarctata.

Backdune Flora

Plants which thrive on the broad dune plain and backdune grow together into dense patches termed *dune mats* that hold the dune together. Vegetation typical of the plain and backdune include Hudsonia tomentosa, Spartina patens, Iva imbricata, and Eregeron glaucus.

Introduced species can outcompete native plants and disrupt animal life, making them formally "invasive species".

Shrub Stage

The above species are herbaceous plants. After they have rooted and developed fully, a second stage, the "shrub stage", can begin. During this phase, larger plants with deeper root systems can be planted. Examples are Empetrum nigrum, Ilex vomitoria, and Vaccinium ovatum. The shrub stage is usually the final phase and may last for short or long periods of time depending on microclimatic conditions such as distance from the shoreline, availability of groundwater, or salt spray effects..

Coastal Management

A single beach may be divided into segments with different owners. Gaining agreement among all owners complicates the process of stabilizing the dunes. Without agreement, some parts of the dune may go unplanted, while others host visually appealing plants that do little to stabilize the dune. Inconsistent vegetation, known as a fragmented or decoupled gradient, can create weak points in the dune that limit its effectiveness against floods and even its continued existence.

Publicly owned beaches, found in U.S. states such as California and Hawaii and in other jurisdictions, present the opportunity to systematically manage—or mismanage—beaches and their accompanying dunes.

Coastal Sand Dune Management

Coastal sand dunes fringe thousands of Kilometres of coastline around the world. They are made up of continuous, hummocky hills of sand that are held together by specially adapted sand dune vegetation. These plants have adaptations that allow them to survive in sand dune habitats, these include adaptations to a moving substrate which covers and uncovers its roots, adaptation to desiccation, to strong winds carrying salt and sand and to wildly changing temperatures. Coastal sand dunes formed during a time of low sea level where sand and sediments on the foreshore became exposed and in conjunction with sufficient winds, this sand was transported up the beach face via a process known as saltation. Once sufficient wind (>15 km/hr) mobilises sand particles they become trapped in vegetation or drift wood at the back of the beach, sand then continues to accumulate among the vegetation and over time slowly builds up the dunes we see today. Sand dunes serve an important purpose by protecting inland areas from coastal water intrusion. They are able to absorb the impact and protect inland areas from high energy storms and act as a resilient barrier to the destructive forces of wind and waves.

Human Impacts

It is generally observed that coastal dunes are highly sensitive to any number of human activities. It is also understood that coastal dunes need to have limited interaction with humans in order for their protection and survival. Around the world coastal dunes are eroding and under pressure due to increased sediment loss because of increased coastal human activities. For example, in the last 30 years, mainly because of tourism, nearly 75% of the Mediterranean's coastal dunes have been damaged or destroyed. Human impacts can expand from minor to highly significant adverse effects. Minor effects include pedestrian trampling, constructing footpaths, and off road vehicle tracks, to the highly more significant which includes the construction of roads, car parks, houses, and the use of sand dunes for ground water extraction, the conversion of sand dunes to land for farming, forestry, waste disposal and sand dune mining.

New Zealand is a country with a very long coast line that consists of around 1090 km of sand dunes. From the early 1900s to the early 2000s the area of sand dunes has declined by about 70% from 129,000 ha to 39,000 ha. This decline has been mainly due to erosion caused by the depletion of vegetation cover in order to convert sand dunes to land for forestry, farming and urban areas. Loss in vegetation cover exposes sand to the action of wind and leaves sand dunes vulnerable to increased erosion as it is vegetation that binds sand together. The loss of vegetation causes sand to move inland which is a process known as sand encroachment, this process can having damaging effects on inland properties and leaves properties vulnerable to future coastal hazards for example storm surge.

As well as human activities, natural processes operate within coastal dune environments, these processes include events such as storms, floods, sea level change, and sediment supply changes. Coastal dune systems are mainly controlled by a principle natural process called the sediment supply and transport system. This system consists of an offshore zone (sediment bank), transit zone (beach and active foredune) and a resting zone (stable sand dunes). The sediment in this system is supplied by long shore drift predominantly from rivers and from the erosion of cliffs, headlands, other dune systems and when sediment supply is low sand dunes are subject to erosion. Humans affect sediment supply in a number of ways including the damming of rivers, dredging in coastal water and the construction of harbours, all of which depletes sediment supply. They also disrupt the long shore transport of sediment through the construction of piers, marinas or groynes, breakwaters and other hard structures. Excavating, building roads, houses and walkways on coastal dunes or beaches also causes sand dune erosion by disrupting natural sediment processes. Effective management of coastal dune systems must consider natural processes as a normal part of the natural environment as well as controlling human activities and seeking to alter natural processes should be kept at a minimum in order for restoration to be successful.

Integrated Management

Sand dunes are an important part of the coast as they are land's first defence against the forces of the ocean. Because of this, management of sand dunes will focus on the natural authenticity and values of the dunes themselves including flora, fauna, and the ecosystem as a whole. The concept of integrated management can be incorporated into management of coastal sand dunes. This concept is made up of three main aspects:

- Restore and conserve natural processes
- Create conditions for future natural processes
- Develop types of land use that agree with the above

These concepts seek to incorporate natural processes and human activities within coastal dunes systems. Sand dune management issues are related to the nature of sediment, which has no cohesiveness so the influence of wind and sediment supply is an important factor in determining erosion and deposition of dry sand grains .

Sand dune management consists of restoring natural dune systems and processes with the implementation of "soft" technique which includes the cessation of any activity that adversely effects the sediment supply of the dunes, planting vegetation, fencing off sensitive areas, creating dune walkways, and providing information to the public regarding the importance of maintaining stable sand dunes. The concept of "soft" techniques is in contrast to the widely used "hard" techniques which consist of protecting the coast through the construction of concrete sea walls, revetments, gabions, groynes and offshore break waters. The problem with using these "hard" engineering methods is that they only provide temporary protection against ocean hazards and each "hard" method produces an "end effect" that causes erosion problems to adjacent coastal areas. This is why the implementation of "soft" techniques in dune restoration should be used more widely in coastal protection as they provide natural, long term protection with no adverse effect to adjacent coastal areas. The use of soft techniques can be seen in many coastal protection projects around the world including many in New Zealand.

Restoration and Protection

One of the first things a coastal manager must look for when establishing sand dune management is to ascertain the main cause of sediment loss and from there, make decisions on how to minimise the cause and how to restore sand dunes appropriately. In New Zealand the main cause of dune erosion has been the loss of sand vegetation and this means that sand dune management consists primarily of stabilizing dune systems through the "soft" technique of re planting and protecting native dune vegetation. It is advised that dunes should not be restored with the planting of exotic species and managers should aim to only use species indigenous to an area, originally New Zealand used exotic, fast growing dune vegetation called Marram Grass) to stabilize eroding sand dunes. It was later discovered that this exotic species created very high, steep sand dunes that were increasing susceptible to New Zealand's high storm energy climate, it also became apparent that the native New Zealand dune vegetation were out competed by this introduced species which caused issues biologically. At present there is a large push to eradicate Marram grass as it is now classed as a pest species in New Zealand and at present coastal managers encourage the planting of the two indigenous species, Pingao (Desmoschoenus spiralis) and Spinifex (Spinifex sericeus) as these two species create low angle, strong, stable dunes.

The management of dune restoration projects includes planting vegetation successfully. Merely scattering and leaving seeds over sandy areas is perceived as a potential waste as strong winds can blow seeds away. Planting of nurseries should be conducted in the hollows of any present sand hills or in depressions of sandy areas. Seeds should be planted in moist conditions and should be lightly covered by a layer of sand and large mature grass individuals should be planted in deep holes with the tops of the plants cut short so as to reduce wind resistance.

Planting of sand dune vegetation alone does not entirely restore or protect sand dunes from human impacts. Seedlings need to be protected from human trampling which is done mainly by fencing off newly planted areas. Fencing also acts as another way of trapping and accumulating sand among the dunes and can also act as protection against strong winds. Figure two shows an area of newly planted vegetation at Spencer Park Beach, New Zealand, it shows that fencing protects plants from trampling and figure three shows how small fencing using chicken wire can encourage plants to grow. In addition to erecting fences to protect sand dune vegetation, the establishment of walkways will protect developed sand dunes from degrading again overtime. These walkways provide access to beaches for the public that does not require pedestrians to cross the dunes themselves thus preventing damage. Figure four shows a typical walkway at New Brighton Beach, New Zealand, the walk way itself is built to go over the sand dunes as opposed to straight through them, in order to maintain the continuity and strength of the dunes.

The Public

Because sand dunes and beaches are extremely popular for outdoor recreation, the success of restoration and dune management depends a great deal on the information available and understanding of the public. This is important because not all aspects of sand dune management are obvious or important to the public if they do not understand why management is necessary e.g. stabilization. There are many ways in which information can be provided to the public including the erection of panels at entrances to dune walkways, leaflets, brochures, maps and council workers who can talk to the public about the importance of protecting sand dunes. Figure five shows a photo of a panel put up at Spencer Park by the Christchurch City Council indicating that people should not walk on the area, as it is an area for sand dune restoration. It needs to be re-iterated that sand dune restoration and protection is in the public's best interest as the loss of sediment on any coast line leads to the erosion of beaches, which in turn leads to a loss of recreational value of many coast lines. The public needs to be aware of the role that sand dunes play in coastal protection because without public support coastal sand dune management will be virtually impossible.

Sand Engine

The sand engine is an experiment in the management of dynamic coastline. It is run off South Holland. A sandbar-shaped peninsula was created by man; the surface is about 1 km^2. It is expected that this sand is then moved over the years by the action of waves, wind and currents along the coast. For the first time in 2011 at the request of the Hoogheemraadschap van Delfland as part of the coastal management and the maintenance of the coastline by the highways and the province of South Holland, a peninsula was created between Ter Heijde and Kijkduin, where natural beaches and dunes are relatively narrow.

Building with Nature

Marcel Stive may be considered as the father of this project. Another such project could provide a solution to the weak coast between Camperduin and Petten, which is called the Hondsbossche seawall. The experience of this first sand engine is interesting for future projects. The construction of the sand engine at Ter Heijde cost 70 million euros and is the first in the world of its kind. Joop Atsma, State Secretary for Infrastructure and Environment, presented the project in November 2011 and his purpose was to convince that the technique of sand engine could be useful on more locations along the Dutch coast.

Delta Dune

The first such project was given the name *Delta Dune*. Work began in January 2011 the conditions were favourable and the operation was completed in October. A volume of 21 million cubic metres of sand covers an area of 128 ha.

Seawall

A seawall (or sea wall) is a form of coastal defence constructed where the sea, and associated coastal processes, impact directly upon the landforms of the coast. The purpose of a seawall is to protect areas of human habitation, conservation and leisure activities from the action of tides and waves. As a seawall is a static feature it will conflict with the dynamic nature of the coast and impede the exchange of sediment between land and sea.

The coast is generally a high-energy, dynamic environment with spatial variations occurring over a wide range of temporal scales. The shoreline is part of the coastal interface which is exposed to a wide range of erosional processes arising from fluvial, aeolian and terrestrial sources, meaning that a combination of denudational processes will work against a seawall. Given the natural forces to which seawalls are constantly subjected, maintenance (and eventually replacement) is an ongoing requirement if they are to provide an effective long term solution.

The many types of seawall in use today reflect both the varying physical forces they are designed to withstand, and location specific aspects, such as: local climate, coastal position, wave regime, and value of landform.

Seawalls are classified as a hard engineering shore based structure used to provide protection and to lessen coastal erosion. However, a range of environmental problems and issues may arise from the construction of a seawall, including disrupting sediment movement and transport patterns, which are discussed in more detail below. Combined with a high construction cost, this has led to an increasing use of other soft engineering coastal management options such as beach replenishment.

Seawalls may be constructed from a variety of materials, most commonly: reinforced concrete, boulders, steel, or gabions. Additional seawall construction materials may include: vinyl, wood, aluminium, fibreglass composite, and with large biodegradable sandbags made of jute and coir. In the UK, *sea wall* also refers to an earthen bank used to create a polder, or a dike construction.

Trade-Offs

A cost benefit approach is an effective way to determine whether a seawall is appropriate and whether the benefits are worth the expense. Besides controlling erosion, consideration must be given to the effects of hardening a shoreline on natural coastal ecosystems and human property or activities. A seawall is a static feature which can conflict with the dynamic nature of the coast and impede the exchange of sediment between land and sea. The table below summarises some positive and negative effects of seawalls which can be used when comparing their effectiveness with other coastal management options, such as beach nourishment.

Advantages and disadvantages of seawalls according to Short (1999)	
Advantages	***Disadvantages***
• Long term solution in comparison to soft beach nourishment. • Effectively minimizes loss of life in extreme events and damage to property caused by erosion. • Can exist longer in high energy environments in comparison to 'soft' engineering methods. • Can be used for recreation and sightseeing. • Forms a hard and strong coastal defence.	• Very expensive to construct. • Can cause beaches to dissipate rendering them useless for beach goers. • Scars the very landscape that they are trying to save and provides an 'eyesore.' • Reflected energy of waves leading to scour at base. • Can disrupt natural shoreline processes and destroy shoreline habitats such as wetlands and intertidal beaches. • Altered sediment transport processes can disrupt sand movement that can lead to increased erosion down drift from the structure.

Generally seawalls can be a successful way to control coastal erosion, but only if they are constructed well and out of materials which can withstand the force of ongoing wave energy. Some understanding is needed of the coastal processes and morphodynamics specific to the seawall location. Seawalls can be very helpful; they can offer a more long term solution than soft engineering options, additionally providing recreation opportunities and protection from extreme events as well as everyday erosion. Extreme natural events expose weaknesses in the performance of seawalls, and analyses of these can lead to future improvements and reassessment.

Simulations

In 2007 researchers at the University of Salerno published studies showing interactions between maritime breakwaters and waves using CAD and CFD software. In the simulations the filtration motion of the fluid within the interstices, which normally exist in a breakwater, is estimated by integrating the relevant RANS equations coupled with a random number generated turbulence model inside the voids, rather than using the classical equations for porous media. The breakwaters were modelled, both for the actual size construction and for a physical laboratory test, by overlapping three-dimensional elements. The numerical grid was thickened in such a way to have some computational nodes along the flow paths among the breakwater's blocks.

Issues

Sea Level Rise: Sea level rise creates an issue for seawalls worldwide as it raises both the mean normal water level and the height of waves during extreme weather events, which the current seawall heights may be unable to cope with (Allan *et al.* 1999). The International Panel on Climate Change (IPCC) (1997) suggested that sea level rise over the next 50 – 100 years will accelerate with a projected increase in global mean sea level of +18 cm by 2050 AD. This data is reinforced by Hannah (1990) who calculated similar statistics including a rise of between +16-19.3 cm throughout 1900–1988. This problem could be overcome by further modelling and determining the extension of height and reinforcement of current seawalls which needs to occur for safety to be ensured in both situations.

Extreme Events

Extreme events also pose a problem as it is not easy for people to predict or imagine the strength of hurricane or storm induced waves compared to normal, expected wave patterns. An extreme event can dissipate hundreds of times more energy than everyday waves, and calculating structures which will stand the force of coastal storms is difficult and, often the outcome can become unaffordable. For example, Omaha Beach seawall in New Zealand was designed to prevent erosion from everyday waves only, and when a storm in 1976 carved out 10m behind the existing seawall the whole structure was destroyed (GeoResources, 2001).

Other Issues

Some Further Issues Include: lack of long term trend data of seawall effects due to a relatively short duration of data records; modelling limitations and comparisons of different projects and their effects being invalid or unequal due to different beach types; materials; currents; and environments (Christchurch City Council, 2009).

Historical Examples

Seawall construction has existed since ancient times. In the 1st century BCE, Romans built a seawall / breakwater at Caesarea Maritima creating an artificial harbor (Sebastos Harbor). The construction used Pozzolana concrete which hardens in contact with sea water. Barges were constructed and filled with the concrete. They were floated into position and sunk. The resulting harbor / breakwater / sea wall is still in existence today - more than 2000 years later.

More recently, sea walls were constructed in 1623 in Canvey Island, UK, when great floods of the Thames estuary occurred, prompting the construction of protection for further events in this flood prone area (Council of Europe, 1999). Since then, seawall design has become more complex and intricate in response to an improvement in materials, technology and an understanding of how coastal processes operate. This section will outline some key case studies of seawalls in chronological order and describe how they have performed in response to tsunami or ongoing natural processes and how effective they were in these situations. Analysing the successes and shortcomings of seawalls during severe natural events allows their weaknesses to be exposed, and areas become visible for future improvement.

Pondicherry

On December 26, 2004, towering waves of the 2004 Indian Ocean earthquake tsunami crashed against India's south-eastern coastline killing thousands. However, the former French colonial enclave of Pondicherry (now Pondicherry) escaped unscathed. This was primarily due to French engineers who had constructed (and maintained) a massive stone seawall during the time which the city was a French colony. This 300 year old seawall effectively kept Pondicherry's historic centre dry even though tsunami waves drove water 24 feet above the normal high-tide mark.

The barrier was initially completed in 1735 and over the years, the French continued to fortify the wall, piling huge boulders along its 1.25 mile (2 km) coastline to stop erosion from the waves pounding the harbour. At its highest, the barrier running along the water's edge reaches about 27 feet above sea level.

The boulders, some weighing up to a ton, are weathered black and brown. The sea wall is inspected every year and whenever gaps appear or the stones sink into the sand, the government adds more boulders to keep it strong (Allsop, 2002).

The Union Territory of Pondicherry recorded some 600 deaths from the huge tsunami waves that struck India's coast after the mammoth underwater earthquake (which measured 9.0 on the moment magnitude scale) off Indonesia, but most of those killed were fishermen who lived in villages beyond the artificial barrier which reinforces the effectiveness of seawalls.

Vancouver

The Vancouver Seawall is a stone seawall constructed around the perimeter of Stanley Park in Vancouver. The seawall was constructed initially as waves created by ships passing through the First Narrows were eroding the area between Prospect Point and Brockton Point. The Vancouver Seawall also exemplifies how seawalls can be utilised and valued for recreational activities and coastal sightseeing. A pedestrian, cycling and roller blading pathway exists on the seawall and has been extended far outside the parameters of Stanley Park.

Construction of the seawall began in 1917, and since then this pathway has become one of the most used features of the park by both locals and tourists and now extends 22 km in total (Belyea & Ross, 1992). The construction of the seawall also provided employment for relief workers during the Great Depression and seamen from HMCS *Discovery* on Deadman's Island who were facing punishment detail in the 1950s (Steele, 1985).

Overall, the Vancouver Seawall is a prime example of how seawalls can simultaneously provide shoreline protection and a source of recreation which enhances human enjoyment of the coastal environment. It also illustrates that although shoreline erosion is a natural process, human activities, interactions with the coast and poorly planned shoreline development projects can accelerate natural erosion rates.

Japan

At least 43 percent of Japan's 29,751 km kilometre coastline is lined with concrete seawalls or other structures designed to protect the country against high waves, typhoons or even tsunamis (New York Times, 2011). When a Tsunami struck in 2011 following a magnitude 9 offshore earthquake, the seawalls in most areas were overwhelmed. In Kamaishi, 4-metre waves surmounted the seawall —the world's largest, erected a few years ago in the city's harbour at a depth of 63 metres, a length of 2 kilometres and a cost of $1.5 billion — and eventually submerged the city centre (Musubi, 2011).

The risks of dependence on seawalls was most evident in the crisis at the Daiichi and Daini nuclear power plants, both located along the coast close to the earthquake zone, as the tsunami washed over walls that were supposed to protect the plants. Arguably, the additional defence provided by the seawalls presented an extra margin of time for citizens to evacuate and also stopped some of the full force of energy which would have caused the wave to climb higher in the backs of coastal valleys. In contrast, the seawalls also acted in a negative way to trap water and delay its retreat.

The failure of the world's largest seawall, which cost $1.5 billion to construct, shows that building stronger sea walls to protect larger areas would have been too costly to be effective. In the case of the ongoing crisis at the nuclear power plants, higher and stronger sea walls should have been built if power plants were to be built at that site. Fundamentally, the devastation in coastal areas and a final death toll predicted to exceed 10,000 could push Japan to redesign its seawalls or consider more effective alternative methods of coastal protection for extreme events. Such hardened coastlines can also provide a false sense of security to property owners and local residents as evident in this situation (Msubi, 2011).

Bibliography

Armytage, W. H. G.: *A Social History of Engineering.* Cambridge: MIT Press, 1966.

Beim, H.; J. Spero and L. Theodore: *Reference Guide for Hazardous Air Pollutants*, Van Nostrand Reinhold, NYC, 1998.

Benjamin Henry: *The Engineering Drawings of Benjamin Henry Latrobe,* Yale University Press, NY, 1980.

Bhavikatti, S. S. : *A Textbook on Elements of Civil Engineering and Engineering Mechanics*, New Age International , Delhi, 2011.

Budin, M. and L. Theodore: *Proceedings of the Mid-Atlantic States Section of AWMA Meeting*, Atlantic City, NJ, 1995.

Chen, W.F.: *Civil Engineering Handbook.* Boca Raton, FL: CRC Press, 1995.

Craig, Ed: *Gas Metal Arc & Flux Cored Welding Parameters*, Weldtrain, Chicago, 1991.

David Fisher: *Rules of Thumb for Engineers and Scientists*, Gulf Pub, Houston, 1991.

Davison, C. S. C. B.: *Engineering Heritage*, Institute of Mechanical Engineers, London, 1963.

Donald Routledge: *A History of Engineering in Classical and Medieval Times*, Open Court Pub., La Salle, 1984.

Drexler, K. Eric: *Engines of Creation: The Coming Era of Nanotechnology*, Doubleday, Canada, 1986.

Emerson, Howard P.: *Origins of Industrial Engineering: The Early Years of a Profession.* Institution of Industrial Engineers, 1988.

Feugerson, E. S.: *Engineering and the Mind's Eye*, MIT Press, Cambridge, 1992.

Fritz Allhoff, Patrick Lin, and Daniel Moore: *What Is Nanotechnology and Why Does It Matter?: From Science to Ethics,* Wiley-Blackwell, Oxford, 2010.

Genod, Jacques: *Fundamentals of Pipeline Engineering*. Houston, TX: Gulf Publishing, 1984.

Gladstones, J.: *Viticulture and Environment*, WineTitles, Adelaide, 1992.

Gregory K. McMillan: *Process/industrial Instruments and Controls Handbook,* McGraw-Hill, New York, 1999.

Hansen, J. R.: *Engineer in Charge: a History of Langley Aeronautical Laboratory, 1917-1958,* NASA, Washington, DC, 1987.

Howard B., Cary,: *Modern Welding Technology*, Pearson Education, Chicago, New Jersey, 2005.

James Brown: *Advanced Machining Technology Handbook,* McGraw-Hill, New York, 1998.

Joseph A. Schetz; Allen E. Fuhs: *Handbook of Fluid Dynamics and Fluid Machinery*, John Wiley & Sons, New York, 1996.

Kalpakjian, Serope: *Manufacturing Engineering and Technology*, Prentice Hall, 2001.

Leone, A.; J. Reynolds and L. Theodore: *Fundamentals of Engineering for Chemical Engineers — Afternoon Exam*, Air and Waste Management Association (AWMA Bookstore), Pittsburgh, PA, 1999.

Louis L.: *Designing Engineers,* MIT Press, Cambridge, 1994.

MacCullough, G.H.: *Elements of Strength of Materials,* Van Nostrand Co., New York, 1957.

McKenna, J.; J. Mycock and L. Theodore: *Handbook of Air Pollution Control and Technology*, CRC Press/Lewis Publishers, Boca Raton, FL, 1995.

Nicholas P.: *Handbook of Chemical Engineering Calculations*, McGraw-Hill, New York, 2004.

Pool, Robert: *Beyond Engineering: How Society Shapes Technology*, Oxford University Press, New York, 1986.

Proffitt, T., R. Bramley, D. Lamb, and E. Winter: *Precision Viticulture: A New Era in Vineyard Management and Wine Production*, WineTitles, Adelaide, 2006.

Punmia, B.C.; Ashok Kr. Jain and Arun Kr. Jain: *Basic Civil Engineering,* Laxmi Publications, Delhi, 2003.

Reynolds, J.; J. Jeris, L. Theodore, *Handbook of Chemical and Environmental Engineering Calculations*, Wiley-Interscience, NYC, 2002.

Richards, E. H.: *Technology and Industrial Efficiency: Proceedings of the Congress of Technology 1911*: McGraw Hill, New York, 1911.

Shigley, J.E.: *Mechanical Engineering Design,* McGraw-Hill, Inc., New York, 1972.

Singh, Gurcharan : *Standard Handbook of Civil Engineering*, Standard Pub, Delhi, 2003.

Sommers, B.J.: *The Geography of Wine: How Landscapes, Cultures, Terroir, and the Weather Make a Good Drop*, Plume Press/Penguin Prentice-Hall Press, 2008.

Stine, Jeffrey K. : *Mixing the Waters: Environment, Politics, and the Building of the Tennessee-Tombigbee Waterway*. University of Akron Press, 1993.

Swamy, H.C.M. : *Elements of Civil Engineering*, Laxmi Publications, Delhi, 2008.

Swinchatt, J., and D.G. Howell: *The Winemaker's Dance: Exploring Terroir in the Napa Valley*, University of California Press, Berkeley, 2004.

Thomas Sixsmith & R. Hanselka: *Handbook of Thermoplastic Piping System Design*. Marcel Dekker Ltd, London, 1997.

Weman, Klas: *Welding Processes Handbook*, CRC Press, New York, 2003.

Wilcox, J. and L. Theodore: *Engineering and Environmental Ethics*, Wiley-Interscience, NYC, 1998.

Index

A

Allmaras Turbulence Model, 252.
Atmospheric Dispersion Modelling, 255, 257.
Automatic Transmission, 114, 115, 116, 117, 118, 119, 120, 121, 122, 123, 124, 125, 126, 127, 128, 129, 130, 131, 132.

B

Balanced Flow, 68.
Beach Nourishment, 196, 267, 271, 272, 274, 287.
Boussinesq Approximation, 50, 51, 56, 57, 58, 61.
Breakwater, 262, 263, 264, 265, 266, 269, 270, 274, 276, 277, 279, 283, 288, 289.

C

Clapotis, 274, 275.

D

Darcy's Law, 50, 53, 64, 65, 66, 67, 68.
Dimensional Analysis, 9, 198, 199, 202, 203, 205, 206, 207, 208, 210, 211, 212, 214, 215.
Dimensionless Concepts, 213.

F

Fluid Dynamics, 2, 7, 10, 38, 40, 41, 42, 45, 54, 56, 58, 60, 66, 89, 90, 92, 94, 95, 280.

H

Hele-Shaw Flow, 50, 53, 54.
High-Density Solids Pump, 136, 137.
Hudson's Equation, 277.
Hydraulic Accumulator, 143, 144, 145, 166, 167, 173.
Hydraulic Brake, 131, 132, 134, 135, 158.
Hydraulic Circuit, 95, 101, 108, 110, 134, 138, 162.
Hydraulic Cylinder, 100, 107, 108, 138, 147, 148, 149, 150, 151, 152, 164, 175, 177.
Hydraulic Engineering, 4, 5, 168, 197.
Hydraulic Fluid, 3, 100, 101, 102, 107, 108, 109, 110, 115, 134, 136, 143, 144, 145, 147, 150, 151, 158, 159, 160, 161, 178.
Hydraulic Machinery, 100, 106, 112, 114, 164.
Hydraulic Motor, 100, 109, 112, 129, 147, 162, 163, 164.
Hydraulic Power Network, 4, 166, 168, 173.

Hydraulic Press, 3, 21, 22, 88, 114, 124, 133, 155, 162, 174, 175, 176.
Hydraulic Ram, 152, 153, 154, 155, 175, 177.
Hydraulic Rescue Tools, 175.
Hydraulic Structure, 7, 240.

I

Iribarren Number, 276, 277.

M

Magnetohydrodynamics, 35, 41, 42, 47, 49.
Mainstream Theories, 218, 220.
Moody Chart, 96, 97.

O

Open-Channel Flow, 97, 88.

P

Percentages and Derivatives, 213.
Pipe Flow, 2, 88, 97.
Pipeline Transport, 223, 228.
Poiseuille Equation, 89, 90.
Position vs Displacement, 205.
Principle of Similitude, 198, 202.

R

Revetment, 263, 264, 266, 269, 270, 278, 279, 280, 283.

S

Sand Dune Stabilization, 280.
Sand Engine, 285, 286.
Seawall, 266, 269, 270, 271, 273, 274, 286, 287, 288, 289, 290, 291.
Slender-body Theory, 50, 54, 56.
Surge Control, 98, 99.

T

Theoretical Physics, 217, 218, 219.
Transcendental Functions, 203.
Turbulence Modelling, 41, 250.

W

Wallasey Great Float, 144.
Wave, 9, 10, 38, 41, 42, 45, 46, 47, 50, 56, 57, 59, 60, 61, 63, 64, 154, 178, 179, 180, 181, 182, 183, 184, 185, 186, 187, 188, 189, 190, 191, 192, 194, 232, 244, 269, 280, 281, 282, 283, 286, 288, 290, 291.

❑❑❑